Hippocampus: Unveiling Multivesicular Release

Nicholas

Table of Contents

Chapter 1

Synaptic Transmission at Central Synapses

1.1 Synaptic dynamics

Information in the brain is communicated between neurons by electro-chemical signalling events at their connection points (synapses). The projections from the neuron are known as dendrites, which are populated by individual presynapses. These individual connection points will be referred to as "single spines" or "single synapses". These events occur when an incoming electrical stimulus surpasses a threshold of response and triggers the exocytotic release of neurotransmitter contained within presynaptic vesicles. In the midbrain, transmission events from single synapses are stochastic and highly variable on a release-by-release basis [1–3]. These events are non-static in nature; in response to repeated presynaptic firing, the postsynaptic response to a given above-threshold stimuli in such synapses will dynamically change with time. On short timescales such effects have been accredited with contributing to such neurological phenomena as working memory, temporal filtering, and cell synchronization [4–6]. At midbrain synapses, synaptic release variability has complicated elucidating the exact mechanics underlying these phenomena. The inability to discern clear mechanics has led to much debate of the underlying properties of the releases, particularly whether multiple vesicles are capable of being simultaneously released from a single synapse in the midbrain [1, 7–12]. In contrast, the neuromuscular junction is well known to release multiple vesicles [13, 14]. A detailed characterization of transmission events is then necessary for understanding information processing in the brain, and how the factors contributing to these events interact and vary through time.

The area of interest in this thesis is the hippocampus; the synapses in this region may also referred to as 'central synapses'. This region is an evolutionarily old and extensively studied part of the brain [15]. Its extensive study makes parsing out the underlying mechanisms of its transmission events critical. That is, the existing studies of this region have, naturally, been formulated upon what was known about information transfer in the hippocampus at the time. If there are details of transmission and underlying synaptic dynamics in this region that have been overlooked in existing models there may be elements missing from our models of the higher order processes in this region. Among higher order processes associated with the hippocampus are spatial memory (which enables navigation), and the conversion from short-term to long-term memory [15, 16]. Addressing these higher order functions necessitates understanding how information is being passed between its cells, and what controls the flow of this process.

Briefly, the overall hippocampal formation can be divided into three major sections: hippocampus proper (Cornu ammonis; (CA) - further subdivided into CA1, CA2 and CA3), dentate gyrus (DG), and subiculum. The entorhinal area (EC) is the main source of synaptic input into the hippocampus, consisting of different layers of neurons that signal to different hippocampal subregions (see Figure 1.1). Within the hippocampus

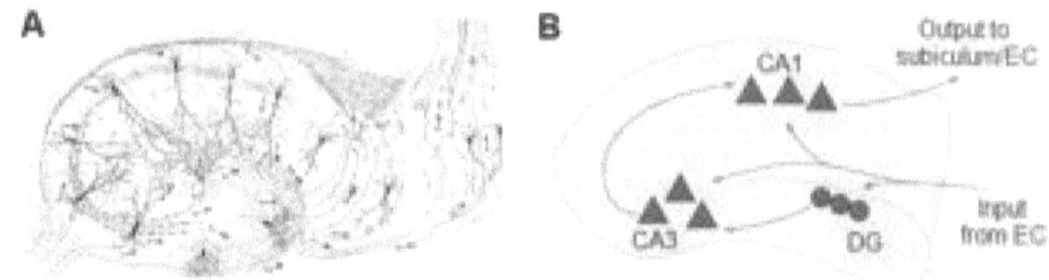

Figure 1.1: **A** A drawing of the hippocampus by Ramon y Cajal [17]; **B** Schematic of the general layout of the hippocampal circuit (Dente gyrus: DG, Entorhinal cortex: EC).

the flow of information is unidirectional: the axons of DG cells, which receive input from the EC, connecting to CA3. In turn, the axons of cells in CA3 relay the signal to their counterparts in CA1. Axons emerging from CA1 then connect to the subiculum or back to the EC. Finally, the subiculum (the main output region of the hippocampus) relays information to, among others, the amygdala, and perirhinal cortex, as well as back to the entorhinal cortex, therefore completing the hippocampal loop [15].

As eluded to, the exact mechanisms underlying transmissions in the hippocampus remains subject to multiple frameworks of description. Three theories have prevailed in the literature: (1) univesicular release, (2) multivesicular release, and (3) the 'kiss and run" theory (the partial exocytosis of vesicles), the last of which is often described in compendium with univesicular or multivesicular release [10,18]. Briefly, univesicular release is something of an oddity when it comes to biological processes, as it suggests that an incoming stimulus will evoke the release of either 0 or 1 vesicles [1,9]. This differs from other neurological regions such as the neuromuscular junction and cerebellum [1]. By constrast, multivesicular release fits much more seamlessly with the existing neurological frameworks. These exocytotic theories will be explored in detail later in this chapter.

Determining accurately which mechanism(s) are present in hippocampal synapses is crucial for elucidating the limits and properties of information transmission in this region. The components underlying transmission characterization, such as the vesicle release framework, can impact energy-cost efficiency and change the upper bound of information able to be transmitted [16,19]. How the characteristics of release change on short time scales with repeated activity will further impact the dynamics of information transfer on the evolving time scales present in neuronal circuitry. Establishing the true mechanisms underlying transmission across synapses is imperative to furthering insight on the dynamics of information flow.

1.1.1 Short-term and homeostatic dynamics

The process of neuronal dynamics changing to adjust for changes in the synaptic efficacy is known as plasticity. On short time scales this is referred to as *short-term plasticity* (STP), a process that is heavily involved in the flow of information, allowing the same axon to communicate independent messages to different post-synaptic targets [5, 20]. Indeed, changes in synaptic transmission have been implicated in underlying higher order functions such as learning and memory [21]. Additionally, STP contributes to the changes in synaptic strength in response to the temporal variability of action potentials on short (10 ms - 10 s) time scales [22–26]. The properties of STP vary widely across projections [27–29], which has lead to the idea that connections can be conceived as belonging to distinct classes [30,31] and information transmission *in vivo* is shaped by these distinct classes [32,33]. To understand information flow and the higher order processes it underlies, characterizing these behaviours correctly is critical.

Short-term dynamics lead to notable changes in synaptic efficacy, however they must not be allowed to do so without regulation, and so are often accompanied by homeostatic plasticity. That is, without a baseline of stable function, the observed changes in synaptic transmission are not meaningful [21]. The primary purpose of homeostatic plasticity is to prevent hyper- or hypoactivity in neuronal circuits [21, 34]. In central neurons where inputs from hundreds of synapses are being integrated it is critical that there are mechanisms in place to prevent over excitation or quiescence. There is a substantial amount of fluctuation in the inputs of these synapses on short time scales, and these fluctuations carry information with them. There has been evidence suggesting homeostatic plasticity in this region acts by controlling firing rates to regulate excitatory synaptic strength [34]. Importantly the effects of homeostatic plasticity have been shown to impact the glutamate release dynamics of hippocampal synapses [35]. This critically relates short-term and homeostatic dynamics, suggesting that both exhibit the ability to regulate information flow at central synapses. Hence, the different varieties of plasticity contribute differently to these higher level processes.

1.2   Prevailing transmission theories

The arrival of an action potential at the presynapse triggers the stochastic release of a synaptic vesicle. This transmission is initiated at the active zone located at the base of the presynaptic terminal where the vesicle pool and proteins necessary for exocytosis are maintained [2, 18]. The vesicles positioned closest to the cell membrane at the active zone are referred to as *primed* or *ready to release* - this subset of the total number of vesicles in the presynapse is referred to as the readily releasable pool (RRP). The remaining un-primed vesicles in the presynapse, located behind the RRP, are a reserve supply that replaces released vesicles in the RRP as transmission occurs. Vesicle size and propensity to release from the presynapse varies across brain regions. Incoming action potentials increase the intracellular calcium concentration due to the voltage gated calcium channels opening, and trigger the initiation of synaptic release events.

This calcium elevation event is known to increase the probability of successful vesicle release in a non-linear power function [36–38]. In the event of a successful release one or more docked vesicles may undergo either partial or complete exocytosis and release their neurotransmitter into the synaptic cleft. The smallest amount of neurotransmitter still capable of transmitting a signal across the cleft is referred to as a quantum, and can be seen as a spontaneous mini-postsynaptic current, however, there is no charge per se at this level. Following release events emptied vesicles are retrieved through endocytosis and refilled to maintain the available supply of vesicles [18,39]. The process of exocytosis and endocytosis is repeated in a cycle with each incoming stimulus to the presynaptic side, and allow the propagation of signals between neurons that underlies all neurological processes.

In the following I describe the prevailing theories for the mechanisms underlying synaptic release at central synapses, delving first into the debated framework for complete exocytosis events as solely univesicular or multivesicular at central synapses. Following this will be discussion of partial transmission, or 'kiss-and-run', events that function as an adjunct to the existing univesicular or multivesicular release frameworks. These descriptions should answer the most pressing questions of: What do they predict? What are their limitations? and How might they be relevant as contributing to information transmission across synapses?.

1.2.1 Univesicular release

In early efforts to understand the sources of variability associated with the release events at central synapses, the two major factors contributing to synaptic releases were found to be the probability of release and the number of vesicles being released. The univesicular framework is a model in which, for a given incoming electrical signal, a maximum of one vesicle will release stochastically across the synapse. Some of the earliest support for this theory came from experimental work done in goldfish Mauthner cells (M cells) in the early 1980's [1, 40]. Most terminations onto M cells have a single release site. The majority of the inhibitory post-synaptic potential (IPSP) recordings from these cells returned a binomial statistic where the number of released quanta was equal to the number of synaptic contacts, supporting a one vesicle per response paradigm. Later physiological studies in hippocampal regions CA1 and CA3 found electron microscopy based evidence of single axonal contacts between pyramidal cells and inhibitory neurons [11, 41]. Performing minimal stimulation excitatory post-synaptic current (EPSC) paired pulse protocols in these regions suggested a zero or one release of vesicles in CA1 [9]. Taken together these results supported univesicular release as the correct model of transmission at central synapses.

The univesicular release (UVR) framework of transmission limits changes in synaptic efficacy to changing as a result of the synapse's release probability or due to changes in the size of the vesicle being released. Increases in the release probability or vesicle size could then comport an increased ability to transmit information. The limitation of synaptic dynamic changes to vesicle size and probability of release places finite constraints on the amount of glutamate that could release. Specifically the limit would be related to the density with which vesicles were packed into the RRP. This upper bound might also limit the maximum amount of information that could be transmitted per release. However, such a limitation could be compensated for to an extent by sparse packing of synapses, allowing for larger vesicles and hence more glutamate to realease.

1.2.2 Multivesicular release

Multivesicular release (MVR) has been well documented as the normal mode of synaptic transmission in the neuromuscular junctions and more recently in the cerebellum [1, 10]. Having the MVR framework as incorrect for the hippocampus then becomes an anomaly relative to the rest of the brain. That said, while the MVR framework was initially rejected as a hypothesis at central synapses [9], it has gained traction in more recent years as a feasible paradigm [8,10,42,43]. Specifically, the high variability observed at central synapses was speculated to result from increased vesicles releasing rather than being entirely a change in release probability [10,42]. MVR describes the process of multiple vesicles being stochastically released from a single synapse in response to an incoming signal. The exact mechanism for this is not yet concretely understood, however, it is believed that the release comes as a product of multiple docked vesicles independently and simultaneously exocytosising through the plasma membrane [10]. Notably, MVR includes the possibility of a single vesicle releasing within its framework, but imposes an upper bound on the releasable number of vesicles only as the available number of docked vesicles. Factors implicated in impacting the number of vesicles released in MVR include: the probability of release, number of vesicles docked, size of the readily releasable pool, and size of vesicles [10,42,44]. Despite central synapses often having low release probabilities there have been instances of MVR found. And further, it has been shown that the quantal content of a single vesicle is unlikely to attain full occupancy of postsynaptic receptors in hippocampal synapses [10,44]. A release of multiple vesicles, however, could attain such an effect by having the neurotransmitter from multiple vesicles interact with a common set of postsynaptic receptors. This increase in postsynaptic occupancy might then be regulated by the presynaptic probability of release, providing MVR a framework for tuning synaptic strength. A variable number of vesicles would also introduce another layer to the overall variability of synaptic transmission, and increase the maximum transmitter amount that can be released. Increasing the amount of neurotransmitter available to be released may also increase the upper bound of information able to be transmitted per event. Collectively this lends itself to the idea that MVR may be a more tractable method of information transmission across central synapses.

1.2.3   Comparing UVR and MVR

The main underlying difference between the multivesicular and univesicular frameworks of release is the ability of central synapses to release more than one vesicle at a time or not (see Table 1.1). While early experimental evidence suggested that UVR was the only occurrence at these synapses, there have since been several instances of experimental evidence for MVR. Further, there are physiological observations that are at odds with the UVR framework as the best description. The quantal content of a single vesicle has been shown to be generally insufficient to saturate postsynaptic receptors [7, 10]. That is, the content of one vesicle is too little to flood all of the postsynaptic receptors such that all binding sites are occupied. This suggests that it would be difficult to increase the strength of a synaptic connection through increased receptor occupancy in a UVR regime. Contrastingly, MVR has been associated with allowing low receptor occupancy synapses to increase their dynamic range [10]. As a consequence of being too small to saturate postsynaptic receptors, UVR releases are also unlikely to fulfill the observed spillover or neurotransmitter rebinding effects that are present at high concentrations of released neurotransmitter [8, 10]. These effects are important as rebinding and spillover affects can play a role in shaping the time course of postsynaptic currents. MVR, having the capacity for a much larger net release, can more easily create such effects. Finally, while UVR has been shown to be almost exclusively present at high release probability synapses [11], MVR has been found at both high and low probability synapses, but with more abundance at low release probability synapses [8, 10]. This result suggests that contrary to early postulations it is not that repeated UVR increases release probability, but rather high release probability decreases the probability of MVR. Put differently, a more reliable synapse may require less release of neurotransmitter to achieve the same effects, and hence regulates a balance between reliability and number of vesicles released. Importantly, this implies MVR to have a role in promoting reliable transmission and changing short-term synaptic dynamics [7,10,42]. Together this provides physiologically observed reasons of why UVR is a less tractable metric for characterizing synaptic transmission.

Type	Full exocytosis	≥ 1 vesicle	Solitary event
UVR	Yes	No	Yes
MVR	Yes	Yes	Yes
KR	No	No	Sometimes

Table 1.1: Summary table of the differences in the major theories of synaptic transmission. UVR: univesicular release, MVR: multivesicular release, and KR: kiss-and-run. Solitary events refer to releases that occur independently of the other categories (e.g. UVR occurring without MVR or KR occurring).

In addition to the behaviour and dynamics based rationals for why UVR is a poor description, there are also theoretical considerations. Notably, synaptic information transfer has been found to generally seek to reduce energetic cost while maximizing information output [19], and to reduce redundancy within the system [45]. UVR being the sole mode of transmission is not conducive to these observations. An example of this discrepancy is in the RRP. For hippocampal synapses the RRP is estimated to contain 4-8 docked vesicles with the reserve pool behind it containing an additional 17-20 vesicles [18]. In the context of UVR, maintaining such a large back supply of vesicles when a maximum of one will release per incoming stimulus seems unnecessarily energetically and spatially costly. Further, a back supply of ≈ 20 vesicles appears excessive for UVR when exocytosed vesicles can be refilled on a millisecond time scale that can keep up with even 200-300 Hz neuronal firing [18]. Having so many vesicles stored would make more sense in terms of energetic cost if multiple vesicles might be released at once. Together this suggests UVR to be an ineffective unilateral framework for transmission.

1.2.4 Partial exocytosis

Having discussed the total exocytosis, or full transmission, models above, we now move on to the case of incomplete or partial exocytosis events. The initial detection of small releases like this was thought to be spillover from nearby transmission events; however, the occurrence of these partial-exocytotic events has since been accepted [2,18]. Known colloquially as "kiss and run" (KR), this theory postulates that in addition to the established full synaptic release events there are micro-releases. That is, release events where the vesicle fuses with the plasma membrane on short time scales ($< 6ms$) and forms a nanometer pore in the vesicular membrane to release neurotransmitter without fully exocytosing [2, 46]. Early detection of these events were found in fusion dye experiments in which the opening of the pore was sufficiently small and fast that neurotransmitter release but not dye release was detected [46]. These releases are referred to as 'running' because the vesicles do not remain attached to the active zone following their release, and recycle rapidly without any clatharin-mediated intermediary [18, 47]. As a result of this, KR has been attributed as a possible method of rapidly replenishing the readily releasable pool [48]. However, as KR still dissociates from the active zone following release, KR has been argued as sub-optimal for fulfilling such a roll. Instead, 'kiss-and-stay' (KS) has been cited as best to optimize the vesicle recycling process. In KS, which is the same as KR in terms of neurotransmitter release, vesicles are reacidified and refilled with neurotransmitters without undocking, thus remaining in the readily releasable pool [18]. This still leaves the question of what role is KR fulfilling? While not unilaterally present KR events are relatively common, having been estimated to occur normally in 20% of vesicles [46]. They occur more often in coincidence with full exocytosis events at low release probability synapses, and have been postulated as a potential mechanism for controlling vesicle size, which gives them an additional role in information transmission [18]. The exact contributions of such small releases to information remains unknown, however the proposed contributions to regulating vesicle size at least implicates the process as being involved in short-term dynamic changes. There are additional considerations in information transfer on whether KR could convey any information by itself, or if there are energetic cost benefits to having such small releases. Together with the described theory for full exocytosis events (summarized in Table 1.1), this provides a complete description of the major theories on release types at hippocampal synapses. These theories then give the framework which models of synaptic transmission are understood by.

Chapter 2

Synaptic transmission and variability at single synapses

2.1   Introduction

Having reviewed the major description frameworks and their properties of vesicle exocytosis at central synapses, we now can begin exploring information communication between neurons. To do this accurately requires a physically tractable model that fully characterizes the properties of transmission events and captures their trial-by-trial variability. Such models are developed to estimate the values of the physically representative parameters involved in transmission; here these include the release probability, the number of vesicles released, and the quantal size [43, 49]. Historically, however, different parameters have been prioritized certain aspects as key contributors to glutamate release variability. Notably models of transmission have often focused on release probability as central in their models [10, 40, 49, 50]. Doing this then negates or minimizes other factors contributing to response variability including vesicle size and number of vesicles released, which may act as potent predictors of response themselves [43]. In earlier models the negation of contributions by the number of vesicles is attributable to the discourse between UVR and MVR frameworks, which lead to many models under-parameterizing transmission events. However, models that do not take into account the variation in vesicle size are then missing an important physical constraint to the release dynamics at these synapses. Modeling transmission events in a physically tractable way is necessary to create a meaningful metric to dissect the ramifications of all the factors underlying the variability of synaptic transmission and hence information transfer.

There have been many proposed models for characterizing release dynamics in the near and distant past, often favouring the use of Gaussian models [3, 10], or Gaussian mixtures [49, 51, 52] with binomial distributions to account for a multivesicular framework of release [9, 40, 49]. There have, however, been attempts made to use skewed distributions [53, 54] and a gamma distribution [55] to fit these events. However, these existing models have not been robust in capturing all of the aspects of transmission with the advent of improved experimental techniques. Recent years have seen an emerging prominence of optical approaches to observing synaptic events, which has given much improved access to the events occurring in synapses. Importantly these methods have allowed for isolation of transmission events at the level of single spines, opening the possibility for quantal analysis [43, 56, 57]. As single quanta do not saturate the post-synaptic side (See Chapter 1) such quantal analysis is critical for understanding the variability in trial-to-trial glutamate release. The improvements in physiological access have now made it possible to improve modeling of transmission to fully capture the dynamics of release events at single synapses.

In this chapter, we show that the choice to neglect contributions of release size in many existing transmission models leads to an incomplete description of transmission. To address this issue, three different approaches to modelling this phenomenon are proposed.

2.2   Transmission model formulation

At most glutamatergic synapses it is now generally agreed on that the variability observed between trials arises primarily from characteristics in the glutamate release profile in the synaptic cleft, and that quantal release events do not saturate the postsynaptic receptors [2,58–60]. As such, creating a model that fits the release profile of glutamate at a quantal level becomes a meaningful metric to investigate the properties of transmission. As stated in the last section, the existing models for synaptic transmission tend to underparameterize their descriptions of transmission. To address this here, we will revisit the underlying factors contributing to synaptic release, and formulate a statistically grounded model that accounts for them.

Visual inspection of the glutamate transients reveals the distributions of response amplitudes to be a right-skewed distribution (Figure 2.1D). This suggested that the prevailing use of Gaussian fit for these distributions, whether mixed or otherwise, may not be the ideal fit. It remains then to find a more appropriate continuous model to fit the release distribution. The bulk of the response profile will be attributable to full exocytosis events; for simplicity, to find a suitable continuous distribution we will restrict our model to the successful releases and take a univesicular framework. These restrictions will then be relaxed as the model is developed. Under these restrictions it is, essentially, the case where release probability is taken as 1 and the number of vesicles is restricted. If we assume that the inner concentration of the vesicles is a steady-state, it is the vesicle size that contributes to differences in the glutamate transient profiles. Having restricted responses in such a manner, it is reasonable to begin by examining the variation in vesicle size.

To begin analyzing vesicle size, electron microscopy data from Qu et al. [61], was used for the mean vesicle diameter and the coefficient of variation. Assuming the diameters to be normally distributed, and subtracting off the average plasma membrane width (2 x $6nm$), these values were used to generate a normal distribution of inner diameters (Figures 2.1A and 2.1B). This distribution can be written formally as:

$$P(d) = \frac{1}{\sqrt{2\pi\sigma^2}}e^{\frac{-(d/2-\mu)^2}{2\sigma^2}}, \tag{2.1}$$

where μ and σ^2 are the mean and variance of the inner diameter, respectively. If the vesicles are assumed to be spherical, their inner volumes can then be found as:

$$v = f(d) = \frac{4\pi(d/2)^3}{3}, \tag{2.2}$$

where d is the inner diameter, and v the inner volume of the vesicle. Having this relationship between diameter and volume, a distribution of inner vesicle volumes was found through a random variable transformation of the form $P(v)dv = P(d)d(d)$. Performing the random variable transformation gives an inner volume distribution of:

$$P(v) = \frac{\sqrt[3]{\frac{2}{\pi}}}{\sqrt{2\pi\sigma^2}\,(3d)^{2/3}}e^{\frac{-\left(\left(\frac{6d}{\pi}\right)^{1/3}-\mu\right)^2}{2\sigma^2}}. \tag{2.3}$$

The resulting distribution of volumes is right skewed, similarly to that of the response amplitudes (Figure 2.1C (Top)). Under the current restrictions placed on the model, and approximating response variability as a function of vesicle size, the distribution of volumes becomes analogous to that of successful response. The Bayesian Inference Criterion was used in previous work [43] to determine the best fit for this skewed distribution, and found to be a gamma distribution (Figure 2.1C (bottom)). Formally, this distribution is:

$$P(v) \approx g\left(v|\gamma_v, \lambda_v\right) = \frac{v^{\gamma_v-1}e^{-v/\lambda_v}}{\lambda_v^{\gamma_v}\Gamma\left(\gamma_v\right)}, \tag{2.4}$$

where the distribution is described by a shape parameter, γ_v, scale parameter λ_v, and mean $\mathbb{E}[v] = \gamma_v \lambda_v$. The subscript v in equation 2.4 references the volume distribution; later references of the parameters without the subscript will be those applying to the synaptic response distribution. That is, γ and λ will parameterize the distributions of univesicular releases. Further, physiologically plausible limits on the ranges of γ and λ have been found for a gamma distribution [43], which are imposed here to aid the physical tractability of the model. Together, the gamma distribution, with biophysically constrained ranges of its parameters, encapsulates the set of univesicular release events in the model, under the assumption of equal glutamate loading across variably sized vesicles.

Having found a suitable continuous distribution to capture the size variability, quantal analysis now becomes more useful. Loosening the restriction of a UVR framework to allowing for MVR, there is then the potential for n vesicles to release in response to an incoming stimulus. Importantly, the addition of two independent gamma-distributed variables results in a random variable that itself is gamma-distributed [43, 55], so the multivesicular framework for successful releases will also appear as a gamma distribution with shape parameter equal to the sum of the shape parameters. While under the right parameterization there could be visibly distinct peaks of the component gamma distributions, it has been shown that within the biophysical range of parameters this is very unlikely [43].

Physically, the n docked vesicles will release independently but simultaneously with a probability, p, which differs from the synapse's overall probability of release, P_r. This release of vesicles can then be described by a binomial distribution:

$$b(k|n, p) = \sum_{k=1}^{n} \frac{n!}{(n-k)!k!} p^k (1-p)^{n-k}, \tag{2.5}$$

where k is the possible number of vesicles released ($1 \leq k \leq n$). Of note, this framework does not specify if or how MVR is distributed in nanodomains. Irrespective of submicron localization when $P_r = 0$, that is when all vesicles have failed to release, a failure distribution will be sampled from. This will be discussed in further detail shortly. The fully successful univesicular framework has now been modified to include both the stochastic probability of releases occurring, p, and the stochasticity introduced by the number of vesicles releasing (n). Letting $\mathcal{V} = P(v)$, the success distribution has then become: $\sum_{k=1}^{n} b(k|n, p)g(\mathcal{V}|k\gamma, \lambda)$.

Finally, a failure distribution is needed to account for both the failed release events and noise considerations for releases. Here this was implemented as a Gaussian distribution with a mean of zero and variance equal to that of the experimental noise, which here was an optical metric $\sigma_{optical}^2$. The optical noise variance used here was found by measuring the variability of glutamate reporter transients when presented with nominally fixed amounts of glutamate by repetitive uncaging [43]. This is not the only way of modeling noise, and in fact, introduces certain biases into the estimate. Specifically, the model is now such that depending on if k is zero the response is drawn from the noise distribution, but the success distribution ($k > 0$) does not contain noise in this formalism, which may bias the parameter estimations somewhat. An alternate method of modeling the noise is to perform a convolution of a Gaussian noise distribution, parameterized by σ_{opt}^2, with each of the component success distributions as well as with a Dirac function at zero for the failures. This will reduce the bias in the estimates, however, is more mathematically complicated. Here, for simplicity, we will proceed initially with the case where the noise is modeled as only occurring in the case of failures, hereafter referred to as the gamma-Gaussian mixture (GGM) model; however, after establishing the simpler case and verifying its utility, I will revisit this alternate, convolutional form of noise in detail. The likelihood function of the GGM model is then:

$$L(\mathcal{V}|\theta) = \prod_{i=1}^{N} b(0|n,p)G\left(\mathcal{V}|0,\sigma_{\text{opt}}^2\right) + \sum_{k=1}^{n} b(k|n,p)g\left(\mathcal{V}|k\gamma,\lambda\right), \tag{2.6}$$

where $G(\cdot)$ is the Gaussian distribution, and the model is described then by a set of five parameters $\theta = \{\sigma_{optical}^2, n, p, \gamma, \lambda\}$. The mean and variance of this model are given by $\mu = np\gamma\lambda$, and $\sigma^2 = \sigma_{opt}^2(1-p)^n + \lambda^2\gamma np(1 + \gamma(1-p))$, respectively. The derivation of the model variance can be found in A. Notably, the maximum number of vesicles released n in this model is a substantial influence on the mean and the variance of this distribution. As the model was developed in a quantal analysis framework, it can be useful to analyze the coefficient of variation (CV) measured in subsected contributions. Doing this gives a metric of examining the relative contribution of our three major distributions (gamma, binomial, and Gaussian) in the model. This can be accomplished using $CV_{UVR}^2 = 1/\gamma$ (gamma) and $CV_{bin}^2 = (1-p)/np$ (binomial); if we then take the ratio $CV^2 = \sigma_x^2/\mu^2$, the CV can be parcelled out as:

$$CV^2 = \frac{\sigma_{opt}^2(1-p)^n}{(np\gamma\lambda)^2} + \frac{1}{np}CV_{UVR}^2 + CV_{bin}^2. \tag{2.7}$$

This expression allows the variability to be discussed in terms of distinct sources, and after values for the parameters are determined gives a metric for evaluating the relative contribution of the different sources. Taken together this completes a statistically based, biophysically tractable model for characterizing synaptic transmission. Having a complete formalism for the model, we will now review briefly the relevant experimental details for the fluorescence data the model will be applied to and subsequently discuss the statistical methods for determining the model parameters.

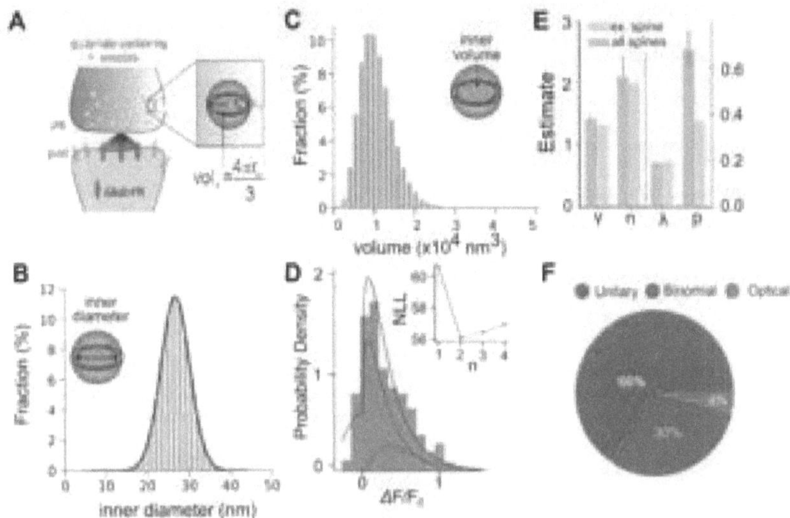

Figure 2.1: **A)** Schematic of transmission and iGluSnFR measurements to infer synaptic vesicle volumes based on assumptions that: *i)* vesicle diameter distribution is uniform and; *ii)* vesicles are roughly spherical. Simulated distribution of **B)** inner vesicle diameters using electron microscopy measurements from Qu et al. (2009) [61]; outer vesicle diameter = 38.7 nm; CV = 0.13; n = 10,000 vesicles. The inner diameter of synaptic vesicles was calculated by subtracting the thickness of the vesicular membrane (2×6 nm). **C)** the inner volumes of vesicles derived from the distribution of diameters presented left. Assuming equal vesicular glutamate concentration, it is expected that the distribution of total vesicular glutamate content mirrors the distribution of vesicle volume. **D)** Evoked fluorescence amplitude histogram for one exemplar spine (gray bars) and probability distribution of the gamma-Gaussian model with properties inferred using the EM algorithm (full red line). Individual release components for $k = 1$ and $k = 2$ are also shown (dashed red lines). Inset: negative log likelihood calculated versus predicted number of vesicles released, n, for the spine shown. **E)** parameter estimates on single stimulation data of spine in **D** and all $N = 18$ spines; error bars are s.e.m **F)** Breakdown of CV contributions by optical noise (optical; green), the stochastic release of 0, 1, or 2 vesicles (binomial; red) and unequal potency of each vesicle (UVR; blue). Panels **A-C** are reproductions of work by C. Soares [43].

2.2.1 Fluorescence measurement of glutamate release

Having now a model formulation to fit the glutamate release profiles, which have been generally agreed to account for response variability, it remains to have data to apply the model framework to. The creation of the model was partially inspired by the advent of improved experimental methods, particularly the contributions of optical protocols in being able to isolate single spine release events. This type of experimental set up was used here by post-doctoral students in Dr. Jean-Claude Béïque's lab, who collaborated with me on this work, through the methods that follow.

As the intention was to study the features of glutamate release, iGluSnFR - a diffuse, plasma membrane-bound optical reporter of glutamate release - was introduced to CA1 neurons along with morphological marker mCherry in organotypic rat hippocampal slices using the interface method originally described in [62] and following protocols approved by the University of Ottawa's Animal Care Committee. This method was used as it allowed for optical signals from single spines to be resolved with high contrast. Hippocampal slices at 6-7 DIV were transfected with gold microparticles coated in cDNA plasmid via gene gun at a ratio of 80/20 by weight of iGluSnFR and mCherry cDNA plasmid, respectively [57]. Imaging experiments were performed 3-5 days after biolistic transfection. Slices were removed from culture and continually perfused with Ringer's solution whilst in a custom recording chamber under an upright microscope (see [43] for details).

Line scans, drawn over a spine and its parent dendrite, were exclusively used to survey the dendritic arbor for responsive spines (Figure 2.2C). Short duration (0.1 ms), low intensity (5-25 mA) stimuli were delivered to the slice at low-frequency (0.1 Hz) while randomly surveying dendritic spines in the apical arbor of transfected cells. To facilitate the process of finding a responsive spine, line scans were performed simulta-neously through multiple nearby dendritic spines and, a paired-pulse stimulus (Figure

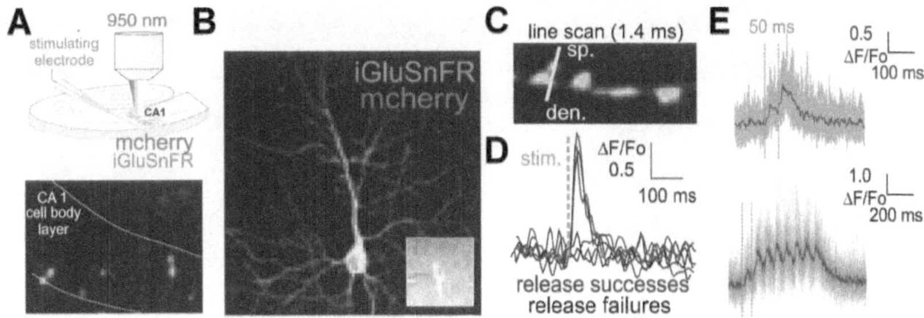

Figure 2.2: **A)** Experimental setup: a glass electrode filled with a fluorescent dye was positioned in the stratum radiatum adjacent to an iGluSnFR-expressing cell and was used to deliver electrical stimuli to the slice to evoke endogenous glutamate release. **B)** Neurons were transfected with both iGluSnFR and mCherry, and expressed variable amount of the fluorescent proteins. Whole-cell recording of an iGluSnFR-expressing CA1 neuron. **C)** Example of line scan set up. **D)** Clearly distinguishable spine successes and failures demonstrate the probabilistic nature of vesicular release at these synapses. **E)** The rapid kinetics of iGluSnFR enables peak-detection at stimulation frequencies that are suitable for studying synaptic facilitation and depression. Reproduction of figures by C. Soares.

2.2E, top) was delivered to increase detection. The sampling frequency of the line-scan experiments was typically in the range of 1.2 - 1.5 ms / line for all optical quantal analysis experiments, which was more than sufficient to fully capture and quantify the rise and falling phases of iGluSnFR transients.

Dendritic spines in the apical dendritic arbor of transfected CA1 neurons were targeted for optical quantal analysis. For evoked stimulation experiments, a glass monopolar electrode filled with Ringer's solution was positioned adjacent to transfected cells in the direction of CA3. Simultaneous two-photon imaging of iGluSnFR and mCherry was performed using a Ti:Sapphire pulsed laser tuned to 950 nm (Figure 2.2A-B). Emission photons were spectrally separated using a dichroic mirror (570 nm) and the emitted light was additionally filtered using two separate bandpass filters (iGluSnFR: 495-540; mCherry: 575-630).

Dendritic spines that were unresponsive to an initial probing phase consisting of 5-10 paired pulse stimuli, were not considered for further analysis, while spines demonstrating responsiveness to these initial probing stimuli were selected for quantal analysis experiments. Before starting an optical quantal analysis experiment at a responsive spine, the stimulus intensity was gradually reduced up to a minimum where time-locked responsiveness was still observed. At successfully responding spines, a single stimulation protocol was performed such that a total of 400 electrical stimuli (40 trials of 10 stimuli) were delivered at varying frequencies (1–8 Hz) while recording the same spine and the peak amplitudes of iGluSnFR events were pooled. Using the same recording set up pairs of stimuli were also delivered at 8 Hz.

As the extracellular calcium level used in the experiments for the single stimuli and paired pulse experiments was high, ≈ 4 mM [43], the same procedure was then performed at a lower, experimentally common calcium concentration, $\approx 2 - 2.5mM$, for trains of stimulations. This was done to attain closer to physiological conditions for the collected data. These trains contained 8 stimuli each and were delivered at 20Hz (Figure 2.2E; bottom) and 8Hz to $N = 7$ and $N = 6$ spines, respectively.

2.3   Data analysis methods

The fluorescence data collected experimentally must then be processed to obtain the response amplitude distribution which the model is applied to. For the initial single stimulation experiments and simulated data, the amplitude of the evoked responses was extracted using a template in a two step process. First, a vector, $\mathbf{k}$, the template time-course, is extracted by computing the trial-averaged fluorescence response triggered by electrical stimulation. This template is then discretized starting at the stimulation time, and ending at a pre-defined time, T, after it. Trial averaging is performed on responses sufficiently isolated in time to be exempt from other synaptic events. For each trial, i, the template is scaled by β chosen to minimize the mean-squared error with the observed fluorescence, $\mathbf{F}_{0:T}^{(i)}$, in the corresponding time window indicated by the subscript $0:T$. Here $\mathbf{F}^{(i)}$ is the matrix with rows containing all trials, $0 \leq i \leq N_{trials}$, of recorded fluorescence response (see Section Fluorescence measurement of glutamate release) from a single spine, and columns for the time course, $0 \leq t \leq T$, of the stimulation(s). The solution to this least-square problem is well known and follows:

$$\beta = \left(\mathbf{k}^T\mathbf{k}\right)^{-1}\mathbf{k}^T\mathbf{F}_{0:T}^{(i)} \tag{2.8}$$

In order to report the maximum of the evoked waveform, β is scaled by the maximum value of the template.

In experiments where multiple simulations were involved, a continuous train of fluorescence with more than one response amplitude is processed. With these repeated stimulations there is often insufficient time between stimuli for the glutamate to be fully cleared and hence fluorescence does not decay to baseline levels before the next stimulation. This necessitates altering the initial method of amplitude extraction to account for contributions from the end of the first stimulus to the beginning of the second and so on (see Appendix B, Figure B.1). As a result, the initial method of extraction was reformulated to take into account the non-zero start point of subsequent stimuli. This was done using multilinear regression, a linear-least-squares method that is useful when a total trace of a mixture is the sum of signals for each component, on each trial, i, in the fluorescence response [63]. The number of stimuli in the trains and their time of delivery was known *a priori* in the data used here. Together this gave an effective method of extracting the amplitudes to yield the response distribution the model is fit to. It remains then to establish a metric to find the parameters for the model.

2.3.1 Statistical inference methods

To find the parameters for the model, we take a statistical approach, making use of variance inference methods. Parameter values have been found using a variety of techniques over the years. These techniques can include Bayesian approaches, such as Markov-Chain Monte Carlo (MCMC) sampling, and non-Bayesian statistics such as least-mean-squares algorithms [64, 65]. These methods are appropriate for parameter searching, however, depending on the framework being considered are not practical. Least mean squares, which is a numerical algorithm designed to find the parameter values that minimize the squared error between the data and model, in particular is prone to finding local minima and does not indicate the uncertainty associated with its estimates [66]. Despite this, least mean squares has been used for the analysis of transmission [67, 68]. While least mean squares has been shown in these cases to, effectively, find parameters that lead to successful fittings of response amplitude data, it has also been shown to lead to inaccurate results.

MCMC algorithms, which have been well studied in the decades they have been in use, begin by making an ergodic Markov chain on a latent variable with a stationary distribution equal to the posterior. The posterior is approximated from an empirical estimate made from the collected samples [65]. This objectively makes MCMC a viable method for approaching parameter inference for the distributions of interest here. However, MCMC algorithms are time inefficient, especially in cases of complex models or large data sets. As a result, *variational inference* - the class of methods used here - has become a preferable alternative for approximating Bayesian inference when complex models or time efficiency is of importance. Variational inference is a method originally from the machine learning literature where probabilities are approximated through optimization [65]. The goal of this approach is to approximate a conditional density of latent variables given observed variables through optimization. Specifically, a family of densities over the latent variables is parameterized by the 'variational parameters'. Through optimization, the setting of the parameters is found as those which minimize the Kullback–Leibler (KL) divergence to the conditional distribution of interest. These fitted variational densities are then analogous for the exact conditional density sought [65]. Using variational inference then allows for an approximation of intractable integrals common to Bayesian inference paradigms. This technique is used to provide an analytical form of the posterior distribution of the unknown variables and perform statistical inference over them, or to derive a lower bound for the marginal likelihood of the observed data. As MCMC algorithms are unfavourable in circumstances of complex models, for the relatively complex model of interest here variational inference methods are a better approach for parameter estimation.

The expectation-maximization (EM) algorithm, which was first described in the late 70's [69], is a statistical method of parameter inference and will be the method used for the model described above. Specifically EM is a method firmly grounded in Bayesian statistics for deriving algorithms to maximize the likelihood function [70]. In this method the estimators end up being dependent on the likelihood function used. This approach has been well documented as being opportune for parameter extraction in mixture models, like the one used here [49,54]. The algorithm works off the principles of gradient descent, iteratively updating estimates of the parameters, θ, and evaluating the likelihood. Letting the F be fluorescence data, the likelihood is:

$$L(F|\theta) = \prod_{i=1}^{N} b(0|n,p)G\left(F_i|0, \sigma_{\text{opt}}^2\right) + \sum_{k=1}^{n} b(k|n,p)g\left(F_i|k\gamma, \lambda\right), \qquad (2.9)$$

Ideally, this will converge to a global maximum; however, convergence is dependent on the landscape of the likelihood function and local maximum may be found instead. As a result, good initial estimates of the parameters are important. The maximization of the likelihood can equivalently be determined as the maximization of the log-likelihood function or the minimization of the negative log-likelihood function.

In the formalization of this method here, we begin with an incomplete set of extracted fluorescence amplitude data, F, which has a likelihood function that is the probability density function, $P_F(f; \theta)$, as a function of θ for a given data set f. Here θ is a vector of unknown parameters belonging to Θ, the set of all valid parameters. A key assumption of this method is the likelihood function has a uniform upper bound, that is $P_F(f; \theta) \leq B < \infty$ for all θ [70]. The maximum likelihood estimator, $\hat{\theta}$, is given by:

$$\hat{\theta} = \underset{\theta \in \Theta}{\operatorname{argmax}} \, P_F(f; \theta). \tag{2.10}$$

While this could be resolved by taking the gradient of the likelihood over the set of parameters in θ and setting each derivative equal to zero, this is not always feasible, such as in nonlinear problems or in cases of large numbers of parameters. The EM method offers an alternative approach. Letting K be a continous random variable with realization k such that $k \in \mathcal{K}$, we call K the latent variable such that,

$$p_F(f; \theta) = \int_{\mathcal{K}} p_F(f, k; \theta) \mathrm{d}k, \tag{2.11}$$

where $p_F(f, k; \theta)$ is the joint probability density function of F and K. Together this pair of variables is referred to as the 'complete data'; that is, the probability density function of F is the marginal of the complete data probability density function over the missing data K. As long as equation 2.11 remains true, the joint probability density function and K may be chosen as complexly or simplistically as the problem dictates.

The major purpose of the expectation step (or E-step) in an EM algorithm is the evaluation of the auxiliary function, $Q\left(\theta; \theta^{(t-1)}\right)$, where $t \geq 0$ indicates an EM iteration with $\theta^{(0)}$ being a valid initial value of the parameter. For steps $t \geq 1$, the auxiliary function is defined as the conditional expectation: $Q\left(\theta;\theta^{(t-1)}\right) = E_{K|F;\theta^{(n-1)}}\left[\log p_F(f,k;\theta)\right]$. For the mixture model used here, this is given by the equation:

$$Q\left(\theta,\theta^{(t)}\right) = \sum_{i=1}^{N} \mu_{i,0}^{(t)} \log\left(b(0|n,p)G\left(F_i|0,\sigma_{\text{opt}}^2\right)\right) \\ + \sum_{k=1}^{n} \mu_{i,k}^{(t)} \log\left(b(k|n,p)g\left(F_i|k\gamma,\lambda\right)\right) \tag{2.12}$$

where $\mu_{i,k}^{(t)} \equiv p_{K|F}\left(k_i = k|F_i,\theta^{(t)}\right)$ is the posterior distribution. Expanded, this gives:

$$\mu_{i,k}^{(t)} = \frac{b\left(k|n,p^{(t)}\right)g\left(F_i|k\gamma^{(t)},\lambda^{(t)}\right)}{L\left(F_i|\theta^{(t)}\right)}. \tag{2.13}$$

The maximization, or M-step, of the algorithm is then for updating parameter estimates such that $\theta^{(t)} = \text{argmax}_{\theta\in\Theta} Q\left(\theta;\theta^{(t-1)}\right)$. The model used here contains three parameters updated during the M-step, $\theta = \{\gamma,\lambda,p\}$. It was possible in this case to arrive at closed form updates of p and λ by taking the first derivative of the likelihood with respect to each:

$$p^{(t)} = \frac{1}{nN}\sum_{i=1}^{N}\sum_{k=1}^{n} k\mu_{i,k}^{(t-1)}, \tag{2.14}$$

and

$$\lambda^{(t)} = \frac{m^{(t-1)}}{\gamma^{(t)}p^{(t)}}, \tag{2.15}$$

where $m^{(t-1)}$ is the model mean given by:

$$m^{(t-1)} = \frac{1}{nN}\sum_{i=1}^{N}\sum_{k=1}^{n} \mu_{i,k}^{(t-1)} F_i \mathcal{H}\left(F_i\right). \tag{2.16}$$

Unlike for λ and p, however, no closed form was possible for γ, and so the maximization of the auxiliary function is used:

$$\gamma^{(t)} = \operatorname*{argmax}_{\gamma} \sum_{i=1}^{N} \sum_{k=1}^{n} \mu_{i,k}^{(t-1)} \log\left(g\left(F_i | k\gamma, \frac{m^{(t-1)}}{\gamma p^{(t)}} \right) \right) \tag{2.17}$$

The model parameter for the number of vesicles, n, was treated as a meta-parameter such that the n which minimizes the negative log-likelihood function was found. The parameter for the optical noise, σ_{opt}^2, was found by fitting a Gaussian noise distribution. Having these equations for the parameter updates, the algorithm iteratively updates the parameters within the valid ranges, Θ, keeping only updated sets of parameters that give a lower value of the negative log-likelihood until a set that minimizes the negative log-likelihood is converged upon.

2.3.2 Simulated data

Before making the final step of data analysis on the fluorescence data, the accuracy of the algorithm was verified on simulated data. To do this N samples were drawn from the model distribution (Equation 2.9) with known parameter values. The EM algorithm was then applied to the simulated data to verify the algorithm's ability to converge upon the true parameter values (see Appendix B, Figure B.2). The bias of the algorithm's estimates goes toward zero in the limit of large sample size ($N_{samples} > 100$). This was repeated for a few sets of parameters, including one which exaggerated the separation between peaks, to verify the viability of the algorithm across data sets. Notably, this magnitude of samples is not often available experimentally, however, the algorithm was found to attain good fits of the simulated data at $N_{samples} \geq 20$. Analysis of the correlation matrix for the parameter estimates showed the number of vesicles, n, to be negatively correlated with the other parameters. This will explain why despite the bias at lower sample sizes, good fits are still attained; the overestimation of one parameter may lead to underestimation of another, however, the obtained fit may be comparable.

Having tested on simulated data and attained good fits, the model and algorithm may be used to investigate synaptic transmission responses.

2.4   Results

Having formulated the model and methods for obtaining estimates of its parameters, we now have a complete metric to examine releases. The model was first applied to amplitude distributions from the fluorescence recordings of single spine stimulations (Number of spines $= 18$) in CA1 (Section Fluorescence measurement of glutamate release). Performing the analysis on single stimulations was done before any repeated stimulation data was analyzed to establish a baseline characterization of the synapses (Figure 2.1E). Having a baseline of the parameters helps evaluate relative change from repeated stimulation, and establishes an intuition of the basic synaptic transmission properties, particularly in comparison to what other models in the literature have described.

In the statistical inference performed here, the single stimulation data demonstrated that, on average, 2 vesicles (n) were present in the stimulations. In the exemplar spine (see Figure 2.1D) this is also the case, and notably in the inset it can be seen that the negative log-likelihood obtained for 3 and 4 vesicles, while not the minimum, were still better fits than 1 vesicle. Taken together this supports that a multivesicular framework is appropriate at central synapses. In addition to considering the estimates of n, the estimates of p were on average slightly higher than might be expected compared to the experimental literature [71, 72], however, this could be a slight selection bias effect, as the spines likely to be chosen for experiments are those which are more responsive. Finally, the parameters of γ and λ do not give very much information in the context of single releases, rather they are useful as a comparison point when multiple stimuli are provided. Together these parameters are now a base of comparison for repeated stimulation conditions, which will then give a metric for seeing what might optimize information transmission.

In addition to providing a baseline parameterization of the synapses, the parameters from the single stimulus experiments can be used to calculate the CV of the model. Hence, having the average parameter estimates, the relative contributions of the different physical aspects of variability used to build the model may be assessed. Noting again the model assumption of steady-state vesicle concentration, it was found that the size of the vesicles (called 'unitary' here as it is the variance of a univesicular release) was the largest contributor to release variance (Figure 2.1F). The binomial aspect of the model (the number of vesicles and their probability of release) contributed second-most, and the optical noise contributed the least. With the unitary aspect providing the most variance, it should be considered that changes in vesicle size may potently contribute to dynamic changes in repeated stimulation conditions. This is particularly important in consideration that existing models of transmission have not considered vesicle size effects [3, 51, 54]. Adding to this the lower contribution of the binomial aspects of the model to the variability the focus of many existing models on release probability as the key determining factor in transmission comes into question. Together this makes an argument for the utility of improved statistical modeling of these transmission events such as the model proposed here.

Importantly, as was mentioned in the formulation of the model, the method used thus far has included optical noise only in the failure distribution. To this point we have shown that the mathematically simpler GGM model will effectively capture the synaptic response distribution (Figure 2.1D). Having established this, it is now interesting to improve on this model by minimizing the noise bias.

2.5 Accounting for noise bias

To review for a moment, the GGM model utilized the optical noise to account for the failure distribution but not any noise effects in the success distribution, potentially biasing parameter estimates. In that formalism, the mixture distribution is such that a latent binomial variable, k, determines whether the fluorescence, F, is drawn from the optical noise distribution, where no vesicles are released, or a noise-free gamma distribution when $k > 0$ vesicles are released. Now an update version of the previously described model is presented to account for the noise bias in the GGM model. In this new formulation of the model, which I will refer to as the noise-corrected gamma-mixture (NCGM) model, a convolution of the Gaussian noise distribution is performed with the successful and failed releases. Specifically, this introduces a convolution into the component gamma distributions for each of the k components ($k = 1, ..., n$). The general form for such a convolution is: $\mathcal{Z}(z) = \int_0^\infty G\left(z - y \mid 0, \sigma^2\right) g(y \mid \gamma, \lambda) dy$, where $\mathcal{Z}(\cdot)$ denotes the density function given by the convolution of Gaussian and gamma densities, $G(z - y \mid 0, \sigma^2)$ and $g(y \mid \gamma, \lambda)$, respectively. Performing this convolution gives:

$$\mathcal{Z}\left(F_i \mid \sigma^2, k\gamma, \lambda\right) = \frac{e^{-\left(F_i/\lambda + \sigma^2/2\lambda^2\right)}}{\lambda^{k\gamma}\Gamma(k\gamma)} \mathbb{E}_{w \mid F_i, \sigma^2, \lambda}\left(w^{k\gamma - 1}\mathcal{H}(w)\right) \tag{2.18}$$

where $w \sim \mathcal{N}(z + \sigma^2/\lambda, \sigma^2)$ and $\mathcal{H}(\cdot)$ is the Heaviside. These changes to the model must then be incorporated into the equations for the EM algorithm. While the likelihood function remains the same, save for replacing $g(F_i \mid k\gamma, \lambda)$ with $\mathcal{Z}(F_i \mid \sigma^2, k\gamma, \lambda)$, the posterior probability, $\mu_{i,k}$ becomes:

$$\mu_{i,k}^{(t)} = \frac{b\left(k \mid n, p^{(t)}\right) \mathcal{Z}\left(F_i \mid \sigma^2, k\gamma^{(t)}, \lambda^{(t)}\right)}{L\left(F_i \mid \theta^{(t)}\right)}. \tag{2.19}$$

Subsequently, the auxiliary equation evaluated in the expectation step of the algorithm is then:

$$Q\left(\theta, \theta^{(t)}\right) = \sum_{i=1}^{N} \left[\mu_{i,0}^{(t)} \log\left(b(0|n,p)G\left(F_i|0,\sigma^2\right)\right) + \right.$$
$$\left. \sum_{k=1}^{n} \mu_{i,k}^{(t)} \log\left(b(k|n,p)\mathcal{Z}\left(F_i|\sigma^2,k\gamma,\lambda\right)\right). \right. \tag{2.20}$$

In addition to updating the auxiliary equation and gamma distribution, the parameter update equations are also adjusted with the convolution. Specifically, the equation for the probability of release (Equation 2.14) remains the same, however, λ can no longer be found in a closed form solution. Instead, γ and λ are found concurrently by maximizing the relevant portion of the auxiliary function with respect to both parameters:

$$\theta^{(t+1)} = \operatorname*{argmax}_{\gamma,\lambda} \sum_{i=1}^{N} \sum_{k=1}^{n} \mu_{i,k}^{(t)} \log\left(\mathcal{Z}\left(F_i|\sigma^2,k\gamma,\lambda\right)\right), \tag{2.21}$$

where the number of vesicles, n, is found as a meta-parameter based on the minimization of the negative log-likelihood as before.

As was done with the GGM model, the NCGM model was tested on simulated data to confirm that parameter estimate bias still tends towards zero for a large sample size. Having verified this, and now having established the formalism for this improved metric of modeling noise, it may be applied to the stimulation data.

2.5.1   Results: noise-corrected gamma mixture model

Beginning by re-analyzing the single stimulation data, an intuition can be gained for how this new metric of modeling noise affects the model. The fits obtained on an exemplar spine by the NCGM model and GGM model are shown in Figure 2.3A. From this it can be seen that, while both models are a reasonable fit to the data, there are also visible differences in their predictions. Performing the analysis on all of the single spine data (see Appendix B, Figure B.3) it was found that convolving the noise with the success distributions led to a significantly higher predicted the shape parameter, γ, (Mann-WhitneyU test; $p \leq 0.0001$) than was obtained using the GGM model, but additionally, the NCGM model had a significantly lower prediction for the scale parameter, λ, (Mann-WhitneyU test; $p \leq 0.0001$), and the probability of release, p,(Mann-WhitneyU test; $p \leq 0.01$); the prediction for n did not significantly differ between the two noise modeling approaches. Although n did not differ significantly between the modeling approaches, its mean value did remain above one. Hence, the MVR framework is still supported when accounting for the bias from the noise in the successful release distribution. Considering those parameters that did change significantly in the NCGM model, the lower predicted value of p is notably closer to the values reported for such synapses in the experimental literature [71, 72]. The other two variables to change significantly, γ and λ, again are not intuitive measures on their own; their mean $(q = \gamma\lambda)$ however, is a more physically tractable term as it is approximately the average vesicle size. In the NCGM model this mean is predicted to be significantly higher than the GGM model (Mann-WhitneyU test; $p \leq 0.01$). Taking all these differences as a complete picture then, the NCGM model suggests synapses to initially have a less regular (lower probability) release of $n \cong 2$ vesicles that are slightly larger than would be predicted if optical noise were allowed to bias the estimates. These observations then update our baseline with which parameters from repeated stimulation experiments will be compared.

To begin characterizing the aforementioned dynamic behaviour that arises with repeated stimulation, we will consider the paired pulse experiments (see Fluorescence measurement of glutamate release). Paired pulse facilitation arises largely due to the notable sensitivity of release probability to extracellular calcium levels that have been reported as $p \propto$ the fourth power of calcium concentration [22], which makes even small changes to the available calcium potent to the release probability. Delivering two stimuli in a small time interval for paired pulse experiments, done here for facilitating synapses, will cause the second stimuli to occur before the extracellular calcium of the first stimuli has been cleared, leading to increased release probability. Having a higher p value will then increase the probability of each individual docked vesicle of releasing on a given stimulus, which ultimately contributes to a stronger release, and hence facilitation event. These effects should be particularly prominent in the conditions here as the paired pulse experiments were conducted at a high concentration of calcium [43] with spines purposefully selected for facilitation as verified by the response amplitude. And indeed, performing the analysis with the NCGM model showed a significant increase in the probability of release, p (Figure 2.3C), although this was the only significant parameter change observed between the pulses from the NCGM model. The number of vesicles (n) and size (q) did not change significantly between the first two stimuli but did both visually trend toward increasing (Figure 2.3C), which is reasonable for facilitating spines, where synaptic strength is expected to increase. Preliminarily, these results suggest that if changes occur to the amount of neurotransmitter being released (n or q) it may take time to occur.

In addition to the results of the NCGM model, Figure 2.3B shows the parameter estimation results for the GGM model for the paired pulse experiments. With this model, the probability of release also increased, although the predicted values were again noticeably higher in the GGM model. Curiously, the GGM model also predicted a significant change in n (Wilcoxon signed-rank test; $p < 0.05$), such that n decreases with repeated stimulation. For a facilitating synapse, this seems to be a counter-intuitive prediction; rather, decreases in n would be anticipated in a depressing synapse. This at-odds prediction seems especially counter-intuitive given these experiments were performed in already high levels of calcium, which would logically exacerbate the effects of facilita-

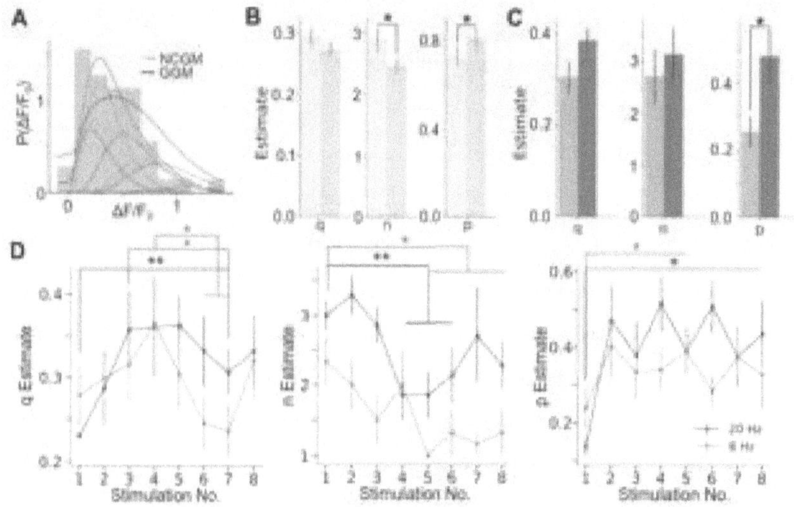

Figure 2.3: **A)** Exemplar fits of the GGM model (black) and NCGM model (blue) applied to fluorescence data from one spine. Dashed lines are components underlying net distribution. Paired pulse estimates for **B)** the GGM model and **C)** NCGM model; **D)** Parameter estimates from 8 Hz (orange) and 20 Hz (blue) trains for the NCGM model. Error bars are s.e.m.

tion. Further, while not significant, the GGM model also has a visual decreasing trend in size, q. As a complete picture, this would indicate the GGM model to predict a sort of stagnancy in the releases where the increased probability of release is compensated for by decreasing the net amount of release, nq, which would not suggest facilitating behaviour. The predictions by the NCGM model then fit much more intuitively with the facilitation literature than those of the GGM form, supporting that modeling the optical noise as included in the success is indeed an improved metric. As such, going forward only the NCGM model will be used.

Proceeding then with the NCGM model, performing analysis on trains of stimuli becomes an important case to look at synaptic facilitation. Specifically, the model will be applied to trains of eight stimuli delivered at $20Hz$ ($N = 7$ spines; Figure

2.2E, bottom). As was mentioned in Fluorescence measurement of glutamate release, the trains experiments were performed at lower concentrations of extracellular calcium ($\approx 2 - 2.5mM$) than the paired pulse or single stimulation experiments, which may be anticipated to impact the resulting characterization somewhat, however, the net facilitation behaviours should remain. These would include: increased probability of release due to residual uncleared extracellular calcium between stimuli, and behaviours conducive to increasing synaptic strength.

Performing the analysis, the probability of release was again significantly increased between the first and second stimuli, and remained significantly higher through the rest of the train relative to the first stimulus (Figure 2.3D; right). This change is consistent with the expectations for facilitating synapses, particularly as the train is at a higher frequency than the paired pulse experiments were conducted. Also similar to what was seen in the paired pulse experiments, the changes in mean size and number of vesicles between the first two stimuli stayed insignificant (Figure 2.3D; left and center). However, as more stimuli are delivered this insignificance does not remain true. Between the first stimuli in the train and the third, there is a significant increase in average size, q, corresponding physically to a predicted increase in vesicle size. This larger size is maintained for the majority of the remaining stimuli. In contrast, between the second stimuli and middle stimuli (four through six), the number of vesicles significantly decreases, before trending toward increasing in number again. This trend of an increased number of vesicles releasing coincides with a trending decrease in the probability of release at the end of the train (Figure 2.3D; right).

The pattern described by the parameter estimates for the trains of stimulation at $20Hz$ meshes with the idea of multivesicular release acts as a type of synaptic strength regulator, controlled by the probability of release as mention in the first chapter [10]. Indeed, as the parametric behaviours for the duration of the train are put into a more complete picture there is an emergent pattern of behaviour. The predictions suggest that repeated stimulation to a facilitating synapse will increase the probability of releasing vesicles, however, after a few stimuli fewer vesicles will release per stimuli. Whether this decrease in the number of vesicles might be the result of a shrinking number of vesicles available to be released, such as many early models of transmission postulated [36], or is the result of a regulatory effect by p remains uncertain. The vesicles releasing are further implicated to be larger, on average, as the number of vesicles releasing decreases (Figure 2.3D; left). This negative correlation between size and number of vesicles may indicate a counterbalance to maintain the release of a certain net amount of neurotransmitter across the synaptic cleft. Alternatively, it could suggest that, as the number of docked vesicles decreases, looser size constraints in the docking area of the presynapse may make having larger vesicles better. Irrespective of the underlying mechanisms, the observed push-pull behaviour observed in this framework suggests a regulatory roll in transmission events where increased synaptic strength is sought after while trying to avoid unnecessary synaptic cleft saturation.

As mentioned, the $20Hz$ trains examined here are both at a higher frequency, and additionally, at a lower level of extracellular Ca^{2+} than the paired pulse experiments were. The first two stimuli of the train exhibiting the same behaviour as the paired pulse experiments lends itself to suggesting that the observed behaviours are, at least initially, robust to changing extracellular $[Ca^{2+}]$. How robust the changes are at lower frequencies of stimulation are then worth examining. Here this is done by analyzing additional eight stimuli trains that were performed at $8Hz$ in the lower calcium conditions. The expectation of this analysis is the behaviour should be approximately consistent with that observed at $20Hz$, if lower in magnitude. And indeed, the probability of release for this data set again significantly increased (Figure 2.3C; right). Additionally, the observable negative correlation between the predicted average size, q, and the number of vesicles, n, remained overall (Figure 2.3C; left and center). However, the trend toward increasing n at the end of the train is much less pronounced at this lower frequency, while the changes in q match more closely to those seen at $20Hz$. This may indicate changes in n to operate on a slower time scale than changes in q or p, allowing saturation to almost be reached before regulation. The maintenance of the trends in parameters between $20Hz$ and $8Hz$ at lower extracellular calcium levels suggests that the predicted behaviours are robust to different frequencies of stimulation.

The changes observed through the use of the NCGM model at varied frequency and duration of stimulation suggest that there may indeed be a mechanistic reason for multivesicular release. Specifically, it may act to increase synaptic strength until such a time as a saturation point is reached, following which univesicularity takes over to allow for strength regulation. In addition to regulatory roles there may exist constraints on energetic cost such that, after releasing several vesicles for a period of time, the cost of replenishment outweighs whatever benefit continuing saturation may have. Whether there are benefits for information transmission in this framework remains to be seen. However, the dynamics described here provide a starting point for describing information dynamics in facilitating synapses.

2.6 Revisiting release probability

While working with the NCGM model, it was noted that the observed response amplitude distributions (see Appendix B, Figure B.4) following synaptic transmission infrequently show any discernible separation between the failure distribution in either noise model (term one of equations 2.20 and 2.12) and the success distribution (term two in equations 2.20 and 2.12). Additionally, examining distributions from similar measures obtained by other groups [50] it became apparent that, while the body of literature on such synapses has long reported bimodal models of the responses, a unimodal fit may function comparably well. That is a framework in which the probability of release for all synapses is $p = 1$. If this is indeed the case, there would be some releases of quite a small size that would normally be encapsulated by the failure distribution, and so be undetected in the bimodal framework of the model. These small release events would fit well with those proposed to occur in the "kiss-and-run" theory described in Chapter 1.

To test this hypothesis, a likelihood ratio test can be used to compare a unimodal form and the bimodal form of the model developed here for fitting the response distribution. The likelihood ratio test is a statistical test for assessing the goodness of fit of two competing statistical models based on the ratio of their likelihoods [73]. In particular, this is a useful statistical metric for comparing when one of the models maximizes over the entire parameter space and the other after imposing some constraint. If the constraint, that is the null hypothesis, is supported by the observed data then the two likelihoods should not differ significantly [73]. Formally the test is evaluated as: $LR = 2(LL(m_2) - LL(m_1))$, where m_2 and m_1 are the two models, and LL is the log-likelihood. For the framework here then, our constraint would be $p = 1$, and the null hypothesis is taken as $\mathcal{H}_0$: '*there are no failures*'. Before this test could be performed, however, the model framework must be updated for the proposed unimodal paradigm.

In developing a unimodal framework for the model, the variable n is purposefully kept to allow the continued biophysical constraint on the ranges of γ and λ that were determined from the electron microscopy data for vesicle size when the GGM model framework was being developed [43] (Section Transmission model formulation). However, this choice effectively makes n a scaling factor to account for the skew in the response distributions, and so it is not rigorously accurate to continue discussing n as a number of vesicles. Keeping with the findings of the previous section, we will also maintain the convolution between the Gaussian distribution to model the noise and the gamma distribution. Notably, the gamma distribution in the unimodal framework is now capturing all releases rather than a subset of 'successful' releases. Together these changes then update the likelihood function such that:

$$L(F_i|\theta) = \sum_{k=1}^{n} \mathcal{Z}\left(F_i|\sigma_{opt}^2, k\gamma, \lambda\right).$$ $$(2.22)$$

Algorithmically, the parameter equation for γ and λ is then performed as the argument maximization of the auxiliary function, as was done in the bimodal NCGM model (Equation 2.21), and n, while no longer the number of vesicles, will still be found as a meta parameter. This then gives the method for finding the likelihood of the unimodal version of the model.

Using the EM algorithm to obtain the parameters for the unimodal case, the likelihoods of both the bimodal and unimodal versions of the model were found for $N = 12$ response distributions. Performing the likelihood ratio test for each of these, it was found that the null hypothesis could not be rejected ($p > 0.05$; for all $N = 12$ response distributions); an example of the fits of the unimodal and bimodal models is shown in Figure 2.4A. From this statistical test result, a unimodal success framework where the synaptic response varies only in size, μ, of release event (where we set $\mu = n\gamma\lambda$) cannot be statistically rejected. Considering this, and having the updated model to capture the dynamics it proposes, it becomes interesting to revisit the data analysis done for the NCGM model in this new unimodal framework.

In reexamining the data, we begin as before with the parameters for the paired pulse protocol (Figure 2.4B). From this, the value of μ increases non-significantly between the first and second stimulation of the protocol. This is approximately in line with what was seen in the NCGM model, as the only significant change in that framework was

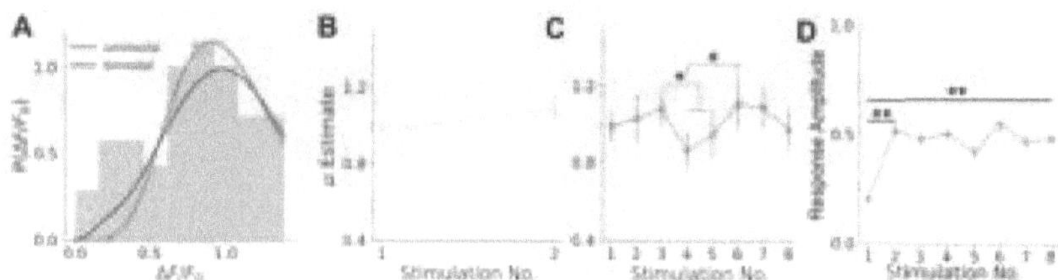

Figure 2.4: **A)** Comparison of unimodal and bimodal (NCGM) fits on fluorescence response amplitude data. Unimodal model parameter estimates for quantal size release ($\mu = n\gamma\lambda$) in **B)** paired pulse experiments, and **C)** $20Hz$ stimulation train. **D)** Extracted response amplitudes from $20Hz$ train. Error bars are s.e.m; significance by MannWhitneyU test.

in p, but there were non-significant increasing trends in q and n (Figure 2.3C). This is reassuring, as it further supports the notion that switching from a bimodal framework to a unimodal one does not significantly change the predicted behavioural dynamics of the synapse.

The non-significant increase between the first and second stimulation using the unimodal framework persists in the $20Hz$ trains (Figure 2.4C). However, in the $20Hz$ trains while the initial three stimulations maintain the trend of non-significant increase, there is a significant decrease in μ at the fourth stimulation of the train; that is μ for the fourth and fifth stimuli is predicted significantly lower than for the third. Following this, there is a significant increase in the estimated μ between the middle (fourth) and late (sixth) stimulations in the train. That is, the average size of release initially trends toward a slight increase, however, this either cannot be maintained due to cost or is not beneficial to maintain, throughout periods of repeated stimulation. However, after dropping for a short amount of time, the average release size recovers close to its original level. Of not the average size predicted then visually, but non-significantly, decreases towards the end of the train. This raises the question of why go back and forth? Taking the regulatory approach discussed briefly earlier in the chapter, this may also be indicative of a mechanism for maintaining a strengthened synaptic response while avoiding saturation. The parametrically predicted behaviour additionally fits well with the average behaviour in the response amplitude over the course of the train (Figure 2.4D). In the $20Hz$ trains, on average ($N = 7$ spines), there appears to be a drop in response amplitude around the fifth stimulus before a net increase in response for the rest of the train, which is most pronounced at the subsequent sixth stimulus. While this averaged amplitude behaviour should not be over-interpreted when coupled with the predicted changes in μ it suggests the middle of the train to be something of a turning point in behaviour. Physiologically considering both the response amplitudes and predicted average size, this would mean after $\approx 3 - 4$ stimulations the synapse is approaching saturation and must, temporarily, scale back to regulate its behaviour.

What this observed behaviour means theoretically for information transfer becomes less clear in this unimodal framework as the literature on this has focused so heavily on these releases as stochastic events, and while the stochasticity between univesicular and multivesicular is somewhat present in this formalism through keeping n, it is less tractably interpreted. The increased mean quantal size in the release initially appears to lend itself to increased information transmission, however, the purpose of the decrease after a few stimulations is less clear, although based on previous work in studying multivesicularity and synaptic dynamics, this could be a correlate of regulating synaptic strength and receptor saturation [10], however, the exact mechanics of such an outcome are unclear. Nevertheless, this initial unimodal framework of synapse-level characterization of transmission becomes more useful when the dynamics of plasticity begin to be examined on a higher level. Having gone through analyzing the behavioural short-term dynamics of synapses, and particularly in light of the potential regulatory effects of said behaviour, it becomes interesting to investigate another type of dynamics.

2.7 Homeostatic dynamics in synaptic transmission

The short-term dynamics we have been examining so far are changes in synaptic strength that, on variable time scales, contribute to changes in synaptic dynamics. Here homeostatic dynamics are now investigated making use of the same frameworks that have been developed thus far. Homeostatic dynamics are a negative-feedback mechanism for regulating the overall activity level of synapses [34, 74], which allow for the restoration of neuron function to an equilibriated level despite any perturbations that may occur. In central neurons, like those studied here, a common experimental protocol of probing functional homeostasis is activity deprivation using tetrodotoxin (TTX), a sodium channel blocker [35, 74]. TTX binds itself to voltage-gated sodium channels and thus inhibits action potential firing. Incubating cultured neurons with TTX will silence population activity, which causes synaptic transmission to respond to compensate. Ultimately this results in enhanced synaptic transmission at all synapses through the increased expression of the ionotropic, transmembrane glutamate receptor, AMPA [34, 74]. This increase in AMPAR expression leads to enhanced response following the removal of the sodium blocker. It is well documented in the literature that homeostatic plasticity, or network silencing, via TTX is expected to yield a stronger postsynaptic response following the removal of the neurotoxin [34]. It has additionally been shown that homeostatic plasticity can influence glutamate release dynamics on short time scales and Hebbian plasticity rules at hippocampal synapses [35]. Here, we use the unimodal model described above to examine changes in synaptic dynamics of facilitation as a result of homeostatic effects. This will make use of paired pulse experiments that were conducted using a TTX application in both single stimuli and paired pulse experiments, which will then be compared to the already analyzed non-TTX, or control (CTL) condition experiments. These paired pulse experiments were conducted at the same frequencies and calcium conditions as those discussed earlier in the thesis.

Using a paired pulse protocol on $N = 9$ spines in each condition (Figure 2.5A), the dynamics of the CTL and TTX conditions may be explored, which here begins with analyzing the properties of the release distributions themselves. The expectation is that the amplitudes of response occurring in the TTX condition should be higher due to the increased synaptic response and, as the spines are facilitating, the second stimuli is again expected to yield a larger amplitude of response due to residual extracellular calcium. And indeed, the amplitude between the first and second stimulus in the paired pulse increases significantly under both conditions. Interestingly, the amplitudes only differ significantly between CTL and TTX at the first stimulation (Figure 2.5B). That is, the first stimulation in the protocol yields a significantly lower amplitude in CTL than TTX, however, the increase in amplitude between stimulation one and two in CTL is large enough that at the second stimulation the difference between CTL and TTX amplitudes is no longer significant.

An additional measure of interest in the release distributions is how the coefficient of variation (CV) changes (Figure 2.5C). Calculating the mean CV at the initial stimulation showed there to be greater variation in the TTX condition than CTL, however, this variation diminished to an insignificant difference between the conditions at the second stimulus. The CV did tend to decrease on average in both TTX and CTL conditions, however not to a significant measure. Of note, when the CV was analyzed in the response distributions of trains (CTL conditions) the decrease in CV was significant at higher frequencies of stimulation and with lower Ca^{2+} levels (see Appendix B, Figure B.5). As facilitating synapses were being targeted, the expectation is that some level of synaptic strengthening occurs in both TTX and CTL conditions [74]. This then begs the question, what is the source of this difference?

Taking the unimodal success framework described prior and using the EM algorithm to extract the parameter values, one finds that the predicted average size of release

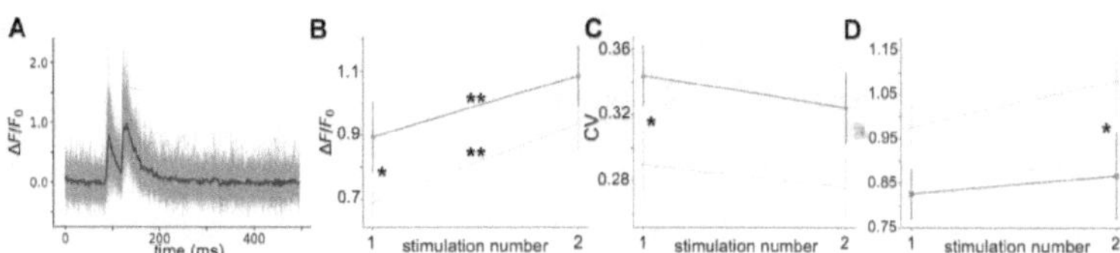

Figure 2.5: **A)** Traces of paired pulse experiment (grey) and average response (blue) under TTX conditions. Control conditions (Yellow) and TTX conditions (Blue) in paired pulse experiments for: **B)** response amplitudes; **C)** coefficient of variation (CV); **D)** average estimate of $\mu = n\gamma\lambda$. Error bars are s.e.m; significance by MannWhitneyU test.

($\mu = n\gamma\lambda$) is larger in CTL conditions at both the first and second stimulation, and the size increases with repeated stimulation in both TTX and CTL conditions (Figure 2.5D). Notably, however, the only significant difference is between the size of release in CTL compared to TTX at the second stimulation (p<0.05; MannwhitneyU test). Putting this together with the changes in the CV and amplitudes of the releases suggests that the larger size and, given the steady-state concentration assumption of the model, increased quantal content of the release in CTL conditions leads to a larger increase in the amplitude of response than that observed in TTX, diminishing the response amplitude gap between them to insignificant. This relationship between quantal content and response amplitude has also been reported previously for paired-pulse experiments at unitary synapses in rat hippocampus under CTL-like conditions [75]. The increase in release size and amplitude is larger in CTL than in TTX, however, the CV decrease is larger in TTX, which may be in part due to the lack of significant change in quantal size released in this condition. That the decrease in CV between stimulations was comparatively small to the change in response amplitude is additionally consistent with EPSP findings in rat motor cortex [76].

Falling in line with past work showing TTX exposed neurons to have a stronger response after blocker removal, the average amplitude observed from TTX conditions was higher. That the increase in μ of TTX is less than in CTL to the extent that they become significantly different, and that TTX is predicted with a lower μ entirely, suggests that the cause of the difference in response amplitude is a result of something other than the presynaptic quantal size. The larger CV decrease in TTX could suggest that the increase in response amplitude induced by TTX is the product of decreased response variation rather than a contribution by the presynaptic quantal size. TTX conditions are known to lead to increased AMPAR content, which would suggest the lower size of release should not be due to any sort of saturation effect. It then remains to ask if the reduced response variation could sufficiently optimize synaptic response to account for the increased amplitude under homeostatic conditions. Of note, there has been evidence for this anti-correlated interaction of amplitude and CV [76, 77]. The more prominent increase under CTL conditions in both response size and amount of amplitude increase could indeed be due to the lack of silencing under such conditions. As was mentioned prior, MVR has been proposed as a mechanism for increasing synaptic strength; without the homeostatic effects from TTX exposure, there will be fewer postsynaptic receptors on average, which may lead to conditions where a larger release from more vesicles would be optimal. As homeostatic plasticity is, effectively, a mechanism for regulating overall activity level these results suggest a reduction of variability and an approximate steady-state in response size as being the regulatory state. The CTL conditions are then the unimpinged facilitation conditions, where the synapse strengths its connectivity through increasing response size.

2.8 Discussion

To review, in this chapter a statistically grounded model of synaptic transmission was introduced based around known biophysical properties. The modeling for this was revised within the chapter resulting in three different iterations: First, as a bimodal model where the noise is modeled as a Gaussian distribution accounting for the release failures (GGM model). A second iteration was created by changing the method of modeling noise by performing the convolution of the Gaussian noise distribution with the gamma-distributed successful releases, leading to a convolved form of the bimodal model (NCGM model). Finally, the NCGM model was reworked into a unimodal form. The main differences between these models are summarized in Table 2.1. Formulations of the EM algorithm for each of these models were applied to response amplitude distributions from stimulation protocols at single spines under different frequency, stimulation, and calcium conditions. Together this gave a method of exploring short-term synaptic dynamics, as well as a brief exploration of homeostatic effects.

Model	Model parameters	Noise in
GGM	σ_{opt}, n, p, q	Failure distrib.
NCGM	σ_{opt}, n, p, q	Failure and success distrib.
Unimodal	σ_{opt}, μ	Success (full) distrib.

Table 2.1: Summary of the model differences and parameters, where $q = \gamma\lambda$ and $\mu = n\gamma\lambda$. GGM: gamma-Gaussian mixture, and NCGM: noise-corrected gamma mixture.

One of the most consistent results across all three models was the notable role of release size in short-term dynamics. The importance of release size to the variation between individual responses was emphasized in the GGM model where the 'unitary' contribution to the variability (Figure 2.1F), was shown to account for approximately two-thirds of the total variability. In the NCGM model the size of releases, q, has an increasing trend even in paired pulse (Figure 2.3C), which then became a significant increase with further stimulation at both 8 and $20Hz$ (Figure 2.3D). Likewise in the unimodal model, there was an increasing trend in the mean size of response for paired pulse (Figure 2.4C), which became significant in the middle and latter half of a stimulation train. These results in both the NCGM and unimodal frameworks support the idea that overall release size may act to regulate synaptic strength [10, 78].

In the homeostatic case (Figure 2.5D) it was seen that the release size did not significantly differ between conditions at the first stimulation; however, the non-significant increase in the control condition with the second stimulation made the control release size significantly larger on the second stimulation than that of the TTX condition. This suggests that the size increases observed in short-term dynamics are not observed in homeostatic plasticity on the same time scales, making changes in size a potent contributor to STP. However, it should be noted that homeostatic conditions were only considered in paired pulse experiments, and may still affect release size on longer time scales with repeated stimulation.

It should be noted that while the models used here are effective for describing the behaviours of release events, exact values should not be assigned based on the parameter estimates. That is, these parameters are useful as a means to describe the net dynamics of vesicle size, number of vesicles released, and release probability (depending on the framework) but should not be interpreted as an exact measurement for these quantities. The values of γ and λ, for example, are based on physiological ranges for vesicle size [43, 61] and provide a mean estimate for the dynamics of vesicle size during STP; however, it would be an overinterpretation to assume an exact vesicle size from this or to take all vesicles released as being the same size. Similarly, direct comparisons between μ in the unimodal framework and q in the bimodal frameworks should not be drawn as q is the mean only of the 'success' (gamma) distribution, whereas μ is the model mean of the unimodal framework. Rather than provide exact values, the models proposed here give a metric of characterizing the net behaviour of synaptic dynamics.

Importantly, the acute impact of release size on the evolving short-term behaviours observed here suggests that the statistical approach to modeling we take is a marked improvement on the characterization of these synapses. This is an aspect that has been missing in many of the existing descriptions for short-term dynamics, which have assigned much more weight to the probability of release [8,11,44]. By taking a statistical approach to modeling here it has been demonstrated that the size of the release itself is important to the characterization of the dynamics, and hence models that have not included such characteristics [10, 40, 49, 50] are effectively underparameterizing the problem. The oversight of such underparameterization is more important still given the statistical acceptance of the unimodal framework, in which case the reliance on release probability as a determining factor in characterizing releases becomes even more ineffective. That said, it would be remiss to assume that statistical acceptance of the unimodal framework indicates it to be the 'true' framework, rather it should be taken as a paradigm that bears further exploration. This is especially true given the unimodal framework's ability to capture similar behavioural dynamics to those determined using the bimodal framework. This will be especially important in future work as accepting a unimodal framework greatly shifts how release events should be characterized, as well as the metric they should be thought of physiologically. Without further rigorous statistical review, the bimodal framework must be considered concurrently with the unimodal framework to ensure there are not any anomalies in the results.

Another consistent result across the frameworks was support for the multivesicular release paradigm at central synapses, as has also been increasingly supported in the literature [7,8,10,42,44]. Considering the number of vesicles releasing, all of the models supported an n with a value greater than one (Figures 2.1E, and 2.3B-D). Notably, in the unimodal model, n is an untenable argument concerning the number of vesicles, as n is kept solely to maintain the biophysical ranges of λ and γ determined using Qu et al.'s electron microscopy data [43, 61] within this framework. Nevertheless, both bimodal models suggest more than one vesicle as the better fit for the early release events analyzed here, which agrees with much of the recent literature on central synapse releases.

Of particular interest, the NCGM model predicted an anti-correlated drop off in n as the value of q increased and vice versa (Figure 2.3D). This is in reasonable agreement with parts of the short-term dynamics literature where it has been proposed that MVR is a synaptic regulatory method to avoid saturation [10]. This builds on the idea that the observed changes in vesicle size underlie a regulatory effect, and that multivesicularity is a framework with which such dynamics can be discussed at central synapses. A regulatory approach to synaptic strength has interesting connotations in information transfer. Indeed the inclusion of release size in the framework may serve as a more tractable metric for parsing out the information transferred across synapses than the release probability dependent forms that have been used prior [19, 49]. Controlling the synaptic strength through the release size such that the postsynaptic receptors do not become saturated may be useful to optimize information transfer across the synapse. That is, there will be more information transferred if the receptors are not simply being saturated on every stimulation beyond the first one or two. Saturating the synapse would be ineffective as no new information would be transmittable after the saturation point was reached, and it would be energetically inefficient to exocytose more neurotransmitter into an already saturated synaptic cleft. Even if the information is calculated maintaining the bimodal view, in which case release probability remains a factor [79, 80], the results shown here strongly support the inclusion of release sizes in such calculations.

Chapter 3

Short-term dynamics in synapses

3.1   Introduction

In the previous chapter, we explored the characterization of synaptic release events at the level of individual spines. Having this description, it now becomes interesting to expand the scope of our analysis to a cellular description. At a cellular level the effects of short-term dynamics as short-term facilitation (STF) or short-term depression (STD), an increasing or decreasing amplitude of response following repeated stimulation respectively, can be more readily explored than at the synaptic level. These diverse effects operate using different time scales, and heavily impact neural dynamics. Short-term dynamics are involved in regulating higher order processes, making the characterization of their behaviours necessary for elucidating a meaningful understanding of these phenomena. Further, the diversity of short-term dynamics suggests a key role of this set of behaviours in controlling information processing [6]. To obtain a characterization of these diverse and evolving dynamics requires a modeling approach that is both able of capturing the experimental behaviour of neurons, but also tractable enough to retain identifiability.

Many existing models, a few of which will be discussed in detail below, approach these dynamics using linear frameworks, capturing the observed behaviours using sets of ordinary differential equations [36, 81–83]. The fitting of these models to data is frequently accomplished using least mean squares approaches, which provide little knowledge of the uncertainty in their parameter estimates and are prone to getting stuck in local minima [66]. While these linear models have been demonstrated to be adequate for fitting the data they were formulated for, they have been seen to be poor fits for the short-term dynamics occurring at cells in other regions, or under altered experimental conditions [6, 36]. While some of this may be a result of insufficient fitting frameworks [66], there are other instances where the model's structure prevents the proper characterization; however, this can be accounted for. That is, with sufficient alteration to the original framework, a model can capture dynamics its original formulation failed to. The inability to capture observed dynamics has sometimes been the result of phenomenological models in linear frameworks being unable to capture occurring nonlinear behaviours, or if they can capture the nonlinearities it is only with sizeable alterations to the modeling framework [6]. Additional complications also arise as phenomenological often use fixed time constants [36, 81–83], thus inhibiting the ability to characterize the multiple time scales present in STP. Taken together, to begin exploring information processing meaningfully requires capturing evolving short-term dynamics, but doing this necessitates a model that reliably fits the diverse, time-scale variant, and nonlinear behaviours observed in short-term dynamics.

A common impasse in modeling synaptic dynamics has been the trade-off that exists between the model's ability to espouse complex experimental data and its identifiability. Many existing models are not favoured because they are either overly complex or restrictive in nature. In contrast, the popularity of some of the linear models, which fail to capture some experimental behaviours, has been attributed to their ease of interpretability. This makes effectively capturing the different types of dynamics observed experimentally a challenging problem; an overly complex model may fit the data well, but will not be useful if it cannot be interpreted. To tackle this problem a mathematically flexible linear-nonlinear model is proposed to efficiently characterize synaptic dynamics. This approach sacrifices some, but not all, of the interpretability to avoid over-parameterization, yet still captures well various plasticity characteristics, a property that has eluded some of the popular existing models.

3.2   Review of short-term dynamics models

Through the last 70 years, there have been several theoretical models proposed to capture the observed short-term dynamics, the earliest of which preceded the discovery of synaptic vesicles in 1955 [36,84]. As with all models, there are advantages and pitfalls to each, and as experimental abilities have improved modifications have been necessary for existing frameworks to adapt to the newly available information. When sizeable modifications are required to account for the new insights into experimental behaviour, it is often preferable to create a new model with a more plausible framework. Here, we will briefly review three popular existing models of STP - the vesicle depletion model [82], the use-dependent vesicle replenishment model [83], and the Tsodyks-Markam model [81] - to answer: How do they work? What assumptions were made in their formulation? And, what are their limitations in light of current knowledge of synaptic transmission? Having this information will then provide a more tractable metric with which to evaluate the effectiveness of the linear-nonlinear framework proposed here that will be developed later in the chapter.

3.2.1 Vesicle depletion model

The vesicle depletion model was first proposed as a model for short-term depression by Liley and North in 1953 [82], before the discovery of synaptic vesicles by De Robertis and Bennett (1955) [84]. As a result of this, the original model proposes a framework where the release of neurotransmitter is described in a rate-based manner rather than a more discrete, vesicle-based framework. Modifications to the original formulation were made to account for both the discovery of vesicles and through further modifications to capture facilitation dynamics. The model assumes transmitter depletion and the associated reduction in postsynaptic response is caused by tetanic stimulation and describes this occurrence through a first-order kinetic model. This model has a replenishment term (first term) and a release term (second term) to determine the occupancy of the release pool, $n(t)$, given as [36]:

$$\frac{dn(t)}{dt} = \frac{1 - n(t)}{\tau_r} - \sum_j \delta(t - t_j) \cdot p \cdot n(t) \tag{3.1}$$

where τ_r is the time constant of replenishment, p is the probability of release, and t_j are presynaptic spike times. This model predicts an exponential decay for the postsynaptic response to constant-rate stimulation. While this was found to be a good fit for some STD dynamics, such as *in vivo*-like EPSCs during stimulation of the calyx of Held [85], it was often ineffective at capturing activity patterns at other synapses, although its mathematical simplicity has allowed this model to remain popular despite its shortcomings concerning more complex STP behaviours. In particular, the depletion model was ineffective at capturing behaviours with slow forms of STD and STF, which has important effects on firing rate modulation on time scales of tens of seconds [36, 86], and have been shown to support differential responses to time-varying stimuli when acting in combination [87]. The depletion model predicts the input frequency to be inversely related to the steady-state level of depression. However, while this was found to fit the depression observed in some synapses well, there are often deviations from this behaviour. Notably, the steady state values decrease more slowly with increasing frequency than is predicted by the model [36]. Additionally, this model does not take into account the contribution to synaptic depression made by the activity dependence of release probability [88] and fails to capture the acceleration in vesicle replenishment following intense stimulation [36]. These studies demonstrate that, while this model is promising due to its mathematical ease, it misses a variety of behaviours that are important to a complete characterization of short-term dynamics.

3.2.2 Use-dependent vesicle replenishment model

This model proposed in 2004 by Furhmann et al. [83] was created to account for the slow reduction in steady-state depression following strong stimuli that the depletion model fails to replicate. This synaptic depression model was targeted at depressing cortical synapses and suggests the vesicle replenishment time constant to be directly modulated by presynaptic activity. This was again proposed using a simple ordinary differential equation [36]:

$$\frac{d\tau_r(t)}{dt} = \frac{\tau_{r0} - \tau_r(t)}{\tau_{FDR}} - a_{FDR}\tau_r(t) \cdot \sum_j \delta(t - t_j) \tag{3.2}$$

where a_{FDR} is a constant to modulate the time constant for each impinging action potential before it recovers to its resting value, τ_{r0}, with a time constant τ_{FDR}. The exact biophysical mechanisms behind use-dependent vesicle replenishment are still not well understood but have been attributed to differences in calcium influx, which may contribute to improved calcium-dependent endocytosis. The ability to transmit across the synaptic cleft during sustained periods of high activity is necessary to explore STP at varying firing rates. A key limitation of this model is that it leaves the enhanced replenishment unbounded and hence will yield a faster decrease of steady-state response amplitude with increasing frequency [36, 83]. As a result, the model will more quickly settle at a constant value which may lead to underestimation of the amount of depression present in some synapses. Additionally, as this model was formulated to account for depression, it lacks the flexibility to be used on facilitation or mixed STF-STD behaviours, so despite its mathematical ease is not robustly useful without further modification.

3.2.3   Tsodyks-Markam model

Finally, we will discuss a prevalent model of short-term dynamics, the Tsodyks-Markam (TM) model. This model has been widely used since its creation in the 1990's [81], largely due to its simplicity and compatibility with observed biological behaviour. Like the previous depletion and use-dependent replenishment models, the TM model was initially derived for modeling short-term depression (STD) in synapses, although it has since been modified to also capture short-term facilitation (STF). In the classic formalism of the TM model, the normalized PSC amplitude, μ_n, caused by spike n of a presynaptic spike train at a given synapse is defined as $\mu_n = R_n u_n$, where the two factors R_n and u_n describe, respectively, the recovered and utilized efficacy of the synapse. The evolution of these variables through time are given by two ordinary differential equations [6]:

$$\frac{dR(t)}{dt} = \frac{1 - R(t)}{\tau_R} - u(t^-)R(t^-)S(t),\tag{3.3}$$

and

$$\frac{du(t)}{dt} = \frac{U - u(t)}{\tau_u} + f\left[1 - u(t^-)\right]S(t); \qquad (3.4)$$

where f is the facilitation constant, $S(t)$ is a spike train, τ_u is the facilitation time scale, U the baseline efficacy, and τ_R the depression timescale. The notation t^- indicates that the function should be evaluated as the limit approaching the spike times from below. The TM model represents facilitation as spike-dependent increases in the utilized efficacy, u. The efficacy increases immediately after each spike and is dependent on the facilitation constant f, and the efficacy immediately before the spike, $u(t^-)$. As a result, increases in u during a spike train will cause the spike-dependent 'jump' to decrease for subsequent spikes. TM models of facilitating synapses are then limited to a logarithmically saturating, or sublinear, facilitation (see Figure 3.2B). This is a problem in this framework as sublinear behaviour is not the only type of facilitation seen in synapses. Rather, improved experimental methods have shown that a supralinear regime emerges in physiological conditions at some synapses. It has been shown recently that adding a small change to the equation for the spike-dependent increase of u will allow for the TM model to capture such supralinear facilitation [6]. In this revised model, given a presynaptic spike train at a constant frequency, the size of the spike-dependent jump increases exponentially for $u < 0.5$ and saturates logarithmically for $u > 0.5$. Thus this modified model allows supralinear facilitation in low efficacy regimes, and switches to sublinear facilitation at larger efficacies (see Figure 3.2C).

While these modifications to the TM model allow more nonlinear dynamics to be captured than the original framework was capable of, it remains that making these sizeable changes to the framework for individual behaviours is not conducive to the widespread use of the model. Moreover, the parameters of the TM model are regularly obtained using least-mean-squares approaches [66], leading to potentially inaccurate estimates of the involved parameters; further, the TM model does not lend itself well to easily fitting the multiple time scales present in synaptic dynamics. As such, while the TM model makes use of biophysically relevant parameters and is easily implemented it, like the two other models discussed here, remains insufficient at capturing all the important aspects of physiologically observed behaviours.

3.3 Linear-nonlinear model: deterministic framework

Having established that the existing models for capturing short-term dynamics are ineffective given the breadth of observable behaviours, we now move to propose a new approach to modeling these behaviours. For simplicity we will begin by modeling the deterministic case of short-term dynamics, which we employ here to capture the behaviour of individual PSC amplitudes, and later build on this framework to account for the stochastic nature of true dynamics. Specifically, we will focus on the TM model which used ordinary differential equation systems to capture the synaptic efficacy separated in several dynamic factors. This method was, as discussed, not entirely effective at capturing the full range of behavioural dynamics. Rather it was found unable to capture observed behaviours without considerable modification to its original formulation. Here we will propose instead a using a linear-nonlinear approach for modeling the synaptic efficacies as a sigmoidal nonlinear readout, f, of a linear filtering operation.

To begin, let $I(t)$ be the post-synaptic current trace comprised of the sum of PSCs at times t_j triggered by incoming presynaptic action potentials from a spike train, and k_{PSC} be the stereotypical PSC time course. The current trace can be written:

$$I(t) = \sum_j \mu_j k_{PSC}(t - t_j),$$
(3.5)

where the relative amplitude, or synaptic efficacy, is now called μ_j for the jth spike in the train, normalized to the first spike in the train ($\mu_1 = 1$). Taking the presynaptic spike train, $S(t)$, as being mathematically described as the sum of Dirac delta-functions, $S(t) = \sum_j \delta(t - t_j)$, the post-synaptic current trace, $I(t)$, can be concisely generated [89]. Here, the objective is to use an efficacy train, $E(t)$, to capture the amplitude changes, and so the time course of the train is taken as invariant, but dynamic amplitudes are assumed for individual PSCs. The efficacy train is made of a weighted sum of Dirac delta-functions: $E(t) = \sum_j \mu_j \delta(t - t_j)$, and can be understood as the multiplication of the spike train and a time-dependent signal, $\mu(t)$ [90]. Doing this the synaptic efficacy is then set at each moment of time.

$$E(t) = \mu(t)S(t). \tag{3.6}$$

With this structure, the current trace can now be taken as a convolution (denoted by $*$) of the efficacy train and the stereotypical PSC shape, $\mathbf{k}_{PSC}$: $I = \mathbf{k}_{PSC} * E$.

It becomes useful now to characterize the synaptic dynamics in terms of the evolving efficacies and presynaptic spikes they respond to. Doing this reflects the typical electro-physiological assays of synaptic properties, which control input spike trains, $S(t)$, and use known PSC shapes ($\mathbf{k}_{PSC}$). As the efficacy is a function of μ and S that depend on time, t, (equation 3.6) this is more easily done by attempting to find the functional $\mu[S(t)]$ mathematically. Using this formalism, a general framework for capturing synaptic efficacy dynamics can be built. Putting this all together the synaptic efficacies can then be reported as:

$$\mu = \frac{1}{f(b)} f\left(\mathbf{k}_\mu * S + b\right), \tag{3.7}$$

where $\mathbf{k}_\mu(t)$ is the *efficacy kernel* describing the spike-triggered change in synaptic efficacy, and b is a baseline parameter, which can be absorbed into the definition of the efficacy kernel, and f is the sigmoidal nonlinear readout. To account for the normalization of amplitudes to the first pulse, the factor $f(b)^{-1}$ is introduced. Parameterizing the efficacy kernel by a linear combination of nonlinear basis functions, $\mathbf{k}_\mu$ can then regulate synaptic efficacy changes after a presynaptic action potential, and $\mathbf{k}_{PSC}$ can regulate the time-course of a single PSC. The efficacy kernel can take any strictly causal form ($k_\mu(t) = 0$ for $t \in -\infty, 0]$), such that a spike at time t_j does not affect the efficacy before or at time t_j. Before taking a sigmoidal readout we now have the 'potential efficacy', the result of the convolution and baseline, $\mathbf{k}_\mu * S + b$. Taken together, changing the shape of the efficacy kernel in the deterministic model presented so far, should allow for the different types of STP to be captured.

To demonstrate this, Figure 3.1A shows a schematic of the model such that a simulated train of four spikes followed by a delay and one more spike are considered. This simulated train is then convolved with $\mathbf{k}_\mu$ where the kernel in this example has a mono-exponential decay. Depending on the amplitude of the kernel the behaviour captured will change. As shown in Figures 3.1B and C, various short-term dynamics are captured through changing the amplitude. When the sigmoidal readout is taken (Figure 3.1D) and sampled at the spike times (Figure 3.1E) the efficacy train is observed. The associated current trace (Figure 3.1F) will display the same short-term dynamics; for example a positive amplitude kernel will lead to facilitating efficacy, and will hence give a facilitating current trace (STF column in Figure 3.1). For the biphasic case, that is STF&STD, a combination of two exponential decays is required to capture the behaviour. When this was done for the case presented in Figure 3.1, the desired behaviour was captured simply by setting the fast component to have a positive amplitude, and the slow component to have a negative one.

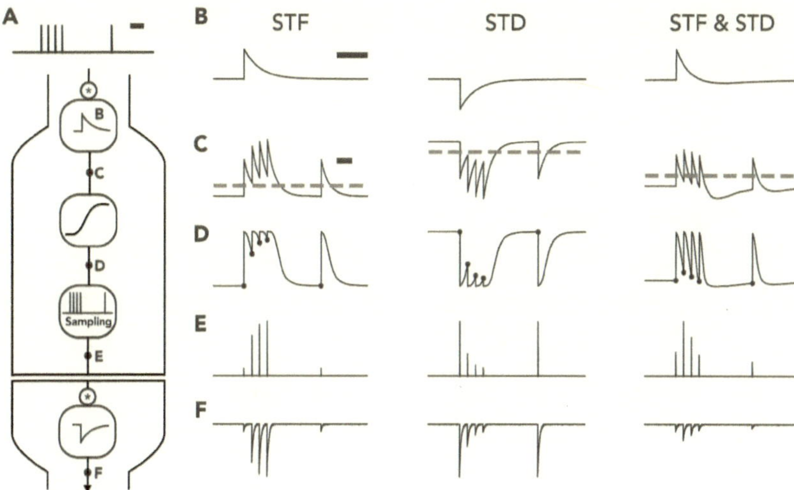

Figure 3.1: Changing the efficacy kernel and baseline in the linear-nonlinear model for short-term dynamics enables capturing STF, and STD dynamics as well as their combination. **A)** The model first passes a pre-synaptic spike train through a convolution with the impulse-response change in efficacy. We illustrate three choices of this efficacy kernel **B)**, a positive kernel for STF (left), a negative kernel for STD (middle) and one for STF followed by STD (right). After the convolution and combination with a baseline (**C)**; dashed line indicates zero), a nonlinear readout is applied, leading to the time-dependent efficacy $\mu(t)$ **D)**. This time-dependent signal is then sampled at the spike times, leading to the efficacy train **E)** and thus to the post-synaptic current trace **F)**. Scale bars correspond to 100 ms. Reproduction of work by R. Naud [6].

3.3.1 Deterministic linear-nonlinear model performance

Having created a deterministic framework for characterizing short-term dynamics, it remains to test the functionality of the proposed model, and compare its performance against existing models for synaptic dynamics. Here this will be against the classic Tsodyks-Markram model described in Section Tsodyks-Markam model. To evaluate the performance of the deterministic framework, we return to the previously described scenario where the TM model was ineffective: capturing supralinear dynamics without sizeable modifications to the original model formulation. Simultaneous activation of multiple synapses may result in sublinear, linear, or supralinear summation depending upon the strength of the stimulus and the location of inputs. A region in which the presence of both supra- and sublinear behaviour has been observed experimentally is in the STF dynamics of hippocampal mossy fiber synapses onto CA3 pyramidal cells (MF-PN synapses) in response to different concentrations of extracellular calcium [6,91]. Data from recent electrophysiology experiments performed by the lab of Dr. Katalin Tóth at these synapses in 2.5 mM and 1.2 mM Ca^{2+} [6], the latter being closer to in vivo conditions and hence more physiologically relevant, is used with permission to explore the effectiveness of the proposed model.

In the electrophysiology experiments, presynaptic spike trains at $50Hz$ were used to evoke strong STF by MF-PN synapses in both Ca^{2+} concentrations (Figure 3.2E). The higher extracellular calcium (2.5 mM Ca^{2+}) is expected to result in a higher probability of release, and hence greater initial EPSC amplitude, and indeed this was observed in the dataset (Figure 3.2A). However, the supralinear facilitation behaviour observed at physiological (1.2 mM Ca^{2+}) is not so intuitive. The cause of the divergence in behaviours between concentrations is associated with the UVR and MVR frameworks discussed in section Prevailing transmission theories. Specifically, MVR is already in place at the higher [Ca^{2+}], and hence recruitment of additional neurotransmitter release sites at the same bouton accounts for all the facilitation observed under this condition [22]. In contrast, STF at low $[Ca^{2+}]$ results from a switch of predominantly univesicular release to the multivesicular paradigm. The glutamate dynamics underlying these two mechanisms at MF-PN synapses are attributable to the complex intra-bouton calcium dynamics in this region [92,93], which lead to gradual and compartmentalized increases in calcium concentration. Consistent with the expectation that these two modes could lie on the opposite sides of the inverse-parabolic relationship between the coefficient of variation (CV) and mean, normal calcium is associated with a gradual decrease of CV through stimulation, while the lower physiological calcium is associated with an increase of CV (Figure. 3.2F). Having such visually obvious differences in behaviour at physiological levels of calcium, it becomes clear that for a model to effectively characterize synaptic dynamics it must be flexible enough to capture both paradigms.

The TM model was shown to effectively characterize sublinear facilitation, but is ill-equipped to replicate the supralinear behaviour observed at low $[Ca^{2+}]$ (Figure 3.2B), which may indicate the original model to have been formulated around higher $[Ca^{2+}]$. The supralinear dynamics can be captured using the TM model, but only after several modifications are made to its original formulation (see Figure 3.2C and section Tsodyks-Markam model). Given the need for such modifications, a framework with less rigidly enforced flexibility is preferable. With that said, the linear-nonlinear model framework proposed here accomplishes just this, switching from sub- to supralinear facilitation (Figure 3.2D) within this framework by keeping the efficacy kernel as it is, but lowering the baseline parameter. The rationale for this is, the baseline parameter can be labelled as the coefficient regulating the amplitude of a constant basis function. If this parameter is high, it is likely that (for a facilitating efficacy kernel) a saturating sublinear part of the nonlinear readout. Conversely, a low baseline parameter will allow the same facilitating efficacy kernel to recruit the onset of the nonlinearity giving rise

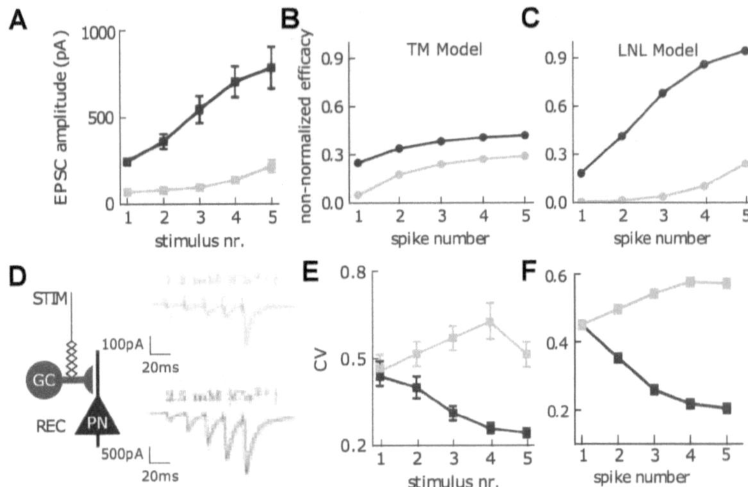

Figure 3.2: **A)** EPSC peak amplitudes as a function of stimulus number. The time course of facilitation varies dependent on the initial release probability. **B)** Synaptic efficacy at each spike according to the classic TM model. **C)** Resulting synaptic efficacy at each spike according to the Linear-Nonlinear (LNL) model. **D)** Mossy fiber short-term facilitation in 1.2 mM (yellow) and 2.5 mM (blue) extracellular $[Ca^{2+}]$. EPSCs recorded from CA3 pyramidal cells (PN) in response to stimulation of presynaptic mossy fibers (GC); $50Hz$, 5 stimuli. **E)** The coefficient of variation (CV), measured as the standard deviation of EPSCs divided by the mean, is increased in 1.2 mM extracellular $[Ca^{2+}]$. Data redrawn from [91]. **F)** The coefficient of variation results from a combination of μ and σ kernel properties. Replicate of work by J. Rossbroich [6].

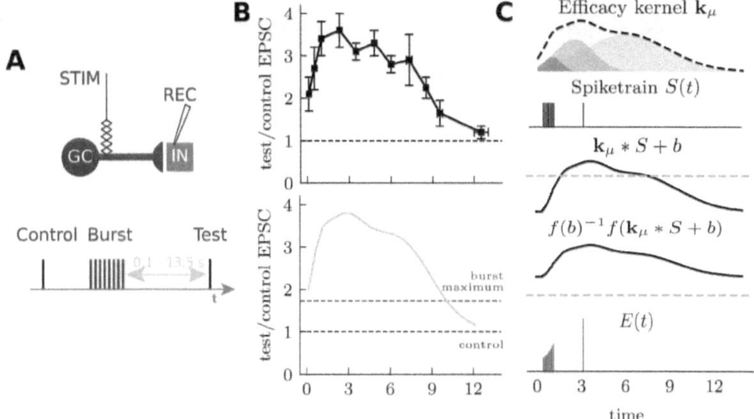

Figure 3.3: **A)** Experimental setup of simulating granule cell (GC) onto interneuron (IN) and **B)** (Top) measurement of post-burst facilitation in CA3 interneurons (redrawn from [23]). (Bottom) Efficacies of test spikes in the synaptic plasticity model as a function of the number of action potentials in the preceding burst. **C)** Synaptic plasticity model. A delayed facilitation kernel was chosen as a sum of three normalized Gaussians with amplitudes {450, 2200, 5500}, means {1.0, 2.5, 6.0} s and standard deviation {0.6, 1.3, 2.8} s. The spiketrain (8 spikes at 100 Hz followed by a test spike) is convolved with the delayed facilitation kernel. A nonlinear (sigmoidal) readout of the filtered spike train leads synaptic efficacies. Dashed lines indicate zero. Replicate of work by J. Rossbroich [6].

to the desired supralinear facilitation.

The efficacy kernel then has high importance in this framework for avoiding the requirement of large structural changes when generalizing to STP's multiple timescales. Indeed changes to sign and timescale of the efficacy kernel have been found to allow for a wide range of STP behaviours to be captured [6], including both STF, STD, and their combination (see Figure 3.1). Taken together the linear-nonlinear model is an improvement on the methods of the existing TM model and is capable of effectively capturing various types of STP through the use of the efficacy kernel.

Having seen that the deterministic formalism of the model can use an efficacy kernel to capture experimentally observed facilitation dynamics caused by changes in extracellular calcium, we now seek to verify its robustness to other experimental behaviours. Particularly, we wish to verify its ability to generalize to multiple timescales and will use the behaviour of mossy fiber synapses onto inhibitory interneurons as an example to show this. These connections have increasing facilitation during the first 2 seconds after a burst of action potentials impinge on them (Figures 3.3A and 3.3B); this is an example of facilitation latency [23] and another behaviour that the classical TM model cannot capture. This is because the TM framework models facilitation as a strictly decaying process, whereas the experimental data shows that facilitation increases during the first 1-2 seconds following the burst. While it is likely that the TM model could capture such dynamics by adding a differential equation for the slow increase in facilitation, this modification consists of a significant modification to the modeling framework, and so a more flexible framework is again preferable.

Unlike the TM model, in the linear-nonlinear framework, the observed facilitation latency can be captured by modifying the shape of the efficacy kernel. An efficacy kernel with a slow upswing, once convolved with a burst of action potential followed by a test-pulse (Figure 3.3C), was found to delay the increase in synaptic efficacy in line with what is observed experimentally (3.3B; bottom) [6]. This demonstrates again the utility of the efficacy kernel in this framework for avoiding the need for robust modifications (such as adding another parameter) to linear frameworks to capture the same behaviours. Thus, provided that the efficacy kernel is parameterized with a basis function spanning a large part of the function space, the proposed linear-nonlinear model can aptly generalize to STP properties unfolding on multiple timescales.

3.3.2 Introducing stochasticity to the framework

While the evidence of the utility of the deterministic framework shown so far has been promising, we must now consider that synaptic transmission is inherently probabilistic. The variability associated with synaptic release depends intricately on the stimulation history, creating a complex heteroskedasticity - the presence of different variability across random variables in response to random disturbances. These dynamics cannot be captured using a deterministic framework, and so we now build upon our existing model to implement a stochastic framework. To begin with we will reformulate the current trace, $I(t)$, such that it can be written in terms of a random variable associated with jth spike and representing a sample of synaptic efficacies, Y_j. The mean will be given by the linear-nonlinear operation:

$$\langle Y_j \rangle \equiv \mu_j = \frac{1}{f(b)} f\left(\mathbf{k}_\mu * S(t_j) + b\right). \tag{3.8}$$

This then lets the current be written as: $I(t) = \sum_j y_j \mathbf{k}_{PSC}(t - t_j)$, where y_j is an instance of Y_j, and the PSCs are taken from randomly chosen amplitudes with an average pattern set by the efficacy kernel. Repeated sampling from the model will produce slightly different current traces, which fits well with the typical outcomes of repeated experimental recordings.

Having a formulation of the current trace within the stochastic framework, a probability distribution is required to capture the synaptic effects. While previous work has argued that a binomial mixture of Gaussian distributions is produced by the quantal release of synaptic vesicles [51,52], it has been demonstrated in the earlier parts of this thesis that releases are better captured by a binomial mixture of gamma distributions. This observation arose as a natural consequence of the assumed Gaussian-distributed inner vesicle diameters and their subsequent cubic transform to obtain the distribution of vesicle volumes (see section Transmission model formulation). In the case of having multiple synaptic contacts, acting under the assumption of equal contribution to the compound PSC by each synapse, a weighted sum of these binomial mixtures will capture the stochastic properties of PSC amplitudes from multiple synaptic contacts [54, 55]. Taken together, this suggests that a skewed distribution is necessary to create a simple parameterization metric for this random process. Such a distribution must additionally have a mean and standard deviation that can change in the course of STP.

Following the work discussed prior in this thesis, the skewed distribution chosen here for the PSCs is the gamma distribution. Specifically, a unimodal gamma distribution is chosen in place of a mixture distribution to maintain the convexity of the function, and in accordance with the statistical validity of using a unimodal distribution discussed in the last chapter (see section Revisiting release probability). Additionally, choosing a solely gamma distribution allows for the convexity of the exponential family distribution to be used. Note that for the linear-nonlinear model explored in this chapter the number of vesicle (n) term used for scaling in the unimodal synaptic transmission model is discarded. For the case here, the distribution is written such that:

$$p(y_j|S(t_j), \theta) = g(y_j|\mu_j, \sigma_j) = \frac{\exp((\frac{-y_i\mu_i}{\sigma_i^2})y_i^{\frac{\mu_i^2}{\sigma_i^2}-1})}{\Gamma(\frac{\mu_i^2}{\sigma_i^2})(\frac{\sigma_i^2}{\mu_i})^{\frac{-\mu_i^2}{\sigma_i^2}}}, \tag{3.9}$$

where the distribution is reparameterized using the mean, $\mu = \gamma\lambda$, and standard deviation, $\sigma = \sqrt{\gamma}\lambda$, of a gamma distribution such that $\gamma = \frac{\mu^2}{\sigma^2}$ and $\lambda = \frac{\sigma^2}{\mu}$. The change from parameterizing with shape and scale to using the mean and standard deviation is useful here as it allows for the characterization of efficacy accomplished for the bimodal model to be maintained. The mean here is set by the linear-nonlinear operation (Equation 3.8) and the standard deviation is set by a possibly distinct linear-nonlinear operation:

$$\sigma_i = \sigma_0 f(k_\sigma * S_i + b_\sigma), \tag{3.10}$$

which we will refer to hereafter as the *variance kernel*, where a baseline parameter, b_σ is introduced, as well as another kernel, k_σ, for controlling the standard deviation. While it could be omitted in the case of standardized data, here the factor σ_0 is introduced to appropriately scale the sigmoidal nonlinearity function f. In this framework, the baseline parameters are fully responsible for the release statistics, as the filters are considered to decay to zero only after a long interval. This allows stationary CV to be obtained as a simple expression; $\text{CV} = \sqrt{1/\sigma_0 f(b_\sigma)}$.

Having developed a stochastic framework, as was the case in our deterministic framework, it remains to verify the ability of this framework to quantitatively capture synaptic dynamics. Unlike in the deterministic approach, however, in the stochastic case there remains a major impediment to characterization: parameter estimation. This is a problem that is also seen in the characterization of cellular dynamics [94], and here requires the implementation of an inference method. As was the case with the synaptic transmission model, a maximum likelihood approach is implemented for inferring the parameters of the linear-nonlinear model, the details of which will be discussed below. Notably, efficient parameter inference depends heavily on the presence of local minima; as a result of this, for the framework here we will investigate the cost function landscape to reaffirm the presence of a minimum.

3.3.3 Parameter inference and fitting

As was discussed in the beginning of this chapter, many of the previous phenomological approaches to modelling STP have taken a least-mean-squares or similar approach to parameter inference. Here, however, we again make use of Bayesian inference techniques and take a maximum likelihood approach to fitting model parameters. In this framework, to infer the parameters, θ, the likelihood equation, $p(\mathbf{y}|S;\theta)$, is established for a given set of amplitudes $y = \{y_1, y_2, .., y_i, ..., y_n\}$ resulting from a stimulation spike-train S. For the formalism used in the linear-nonlinear model, we find the minimum of the negative log-likelihood (NLL). This is given by:

$$NLL(\mathbf{y}|S,\theta) = \sum_i \left[\frac{y_i \mu_i}{\sigma_i^2} - \left(\frac{\mu_i^2}{\sigma_i^2} - 1 \right) \ln \left(\frac{y_i \mu_i}{\sigma_i^2} \right) + \ln \left(\frac{\Gamma\left(\frac{\mu_i^2}{\sigma_i^2}\right) \sigma_i^2}{\mu_i} \right) \right] \tag{3.11}$$

where μ_i and σ_i are shorthand for the efficacy and standard deviation at the ith spike time: $\mu_i = \mu(t_i)$, $\sigma_i = \sigma(t_i)$, the elements of the vectors μ and σ.

The gamma distribution's time-dependent standard deviation and mean are parameterized by expanding the filters $\mathbf{k}_\sigma$ and $\mathbf{k}_\mu$ in a linear combination of nonlinear basis, where the basis functions are chosen here to be exponential functions with different decay times: $k_\sigma(t) = \sum_m c_m h_m(t)$, and $k_\mu(t) = \sum_l a_l h_l(t)$. Here the hyper-parameters are the choice of the number of basis functions, $l \in [0, L]$ and $m \in [0, M]$, and the decay time scale for each basis function $h_l(t) = \Theta(t)e^{-t/\tau_l}/\tau_l$, where $\Theta(t)$ is a Heaviside function. Free parameters are the amplitude of the basis functions $\{a_l\}$, $\{c_m\}$ and the scaling factor σ_0 (from Equation 3.10). By choosing hyper-parameters *a priori*, the number of bases chosen must be sufficiently small to avoid overfitting, yet large enough to avoid model rigidity.

The choice of time constant is made to exhaustively tile the range of physiologically relevant time scales. Of note, the choice of τ does not specify the time scale of synaptic dynamics because a combination of exponential basis functions can be used to capture a decay time scale absent from the set of τ hyper-parameters. The time-scale will be determined by inferring the relative amplitude of the basis functions. The baseline parameter can be labelled as the coefficient regulating the amplitude of a constant basis function, such that $a_0 = bh_0(t) = b_\mu$ and $c_0 = b_\sigma h_0(t) = b_\sigma$. There are thus $L+2+M+2$ free parameters in total :

$$\theta = \{a_0, ..., a_L, \sigma_0, c_0, ..., c_M\}$$

For the case of having a spike train as input, the train is filtered using the set of basis functions, and the resulting filtered spike train just before each spike is stored in a matrix, X. Each individual basis function corresponds to a row of the matrix, and each column corresponds to spike timings. That is, there are as many columns as there are spikes in the train. The matrix, X, thus stores the result of the convolution between the various basis function (rows) at the spike times (columns). For simplicity, it is convenient to take the same choice of basis functions for the efficacy and the variance kernels. The amplitudes are expressed in a vector, $\theta_\mu = \{a_0, ..., a_L\}$, for the efficacy kernel, and $\theta_\sigma = \{c_0, ..., c_M\}$ for the variance kernel. Expressing the linear combination as a matrix multiplication yields:

$$\mu = \frac{1}{f(a_0)} f\left(X^T \theta_\mu\right),$$

and

$$\sigma = \sigma_0 f\left(X^T \theta_\sigma\right),$$

where μ and σ have length n and can be used to evaluate the NLL according to equation 3.11. Performing a grid search of the parameter space around initialized parameter values gives the landscape for the function, which allowed for confirmation of convexity. The inferred parameters will then be the set of θ_μ and θ_σ that minimize the NLL over the training set.

Iteratively varying the shape of the filter time-course the optimal choice is found as the one maximizing the likelihood of synaptic efficacy observations. This is effective due to the automatic characterization method described here using maximum likelihood principles, and the probabilistic nature of this synaptic release model. Additionally, this method offers a few advantages: firstly, the method is firmly grounded in Bayesian statistics, allowing for the inclusion of prior knowledge and the calculation of posterior distribution over the model parameters [55, 66]. And secondly, unlike for prior STP models, the approach described here does not rely on experimental protocols tailored for characterization [36]. However, targeted experiments could be used to improve inference efficiency. Naturalistic spike trains recorded *in vivo* [92, 95], Poisson processes or other synthetic spike trains can be used in experiments to characterize synaptic dynamics in realistic conditions.

3.3.4 Inference on test data

Having now established the maximum likelihood approach to be used to extract parameters, it remains to test the efficiency and limitations of this inference method. However, to do this requires a test set of data. To begin an artificial Poisson spike train is generated for a length of time T (Figure 3.4A). All numerical simulations and parameter inference were implemented in Python using the NumPy and SciPy packages [96, 97]. For simplicity, rather than having filters that are described by a combination of nonlinear basis functions, here a mono-exponential decay kernel, $\mathbf{k}_\mu$, with a known decay time constant is used as a single basis function. In cases where the time constant is unknown, the coefficient of a combination of nonlinear basis functions could be fit, as is typical in other linear-nonlinear models [98–101]. Choosing a known, predetermined set of parameters for the efficacy and variance kernel amplitudes and baselines, as well as for the variance kernel scaling factor, μ and σ are found according to equations 3.3.3 and 3.3.3, respectively. These parameters are used to generate random variable Y from which samples y are drawn, allowing the current trace to be obtained as described in section Introducing stochasticity to the framework. There will still be a small amount of variability in the current trace resulting from the same spike train between given samplings of Y (Figure 3.4B, top).

Taking a long Poisson spike train ($N_{spikes} > 100$) and a known set of parameters for the efficacy and variance kernels, Y and subsequently the current trace (Figure 3.4B, top) were generated through numerical simulation following which the likelihood function was evaluated over a large range of the parameter space. Exploring the results of this evaluation, it was found that the likelihood function appears convex over a large range of parameter values with no observed local minima (contour lines, Figures 3.4C-F), as expected given the chosen exponential family distribution. The slanted elongation of likelihood contours in these plots is indicative of correlations or anti-correlations between parameter estimates. Unsurprisingly, the estimates of baseline and scale factor of the σ-kernel are anti-correlated (Figure 3.4D); conversely, the estimates of filter amplitudes for efficacy and variance have little correlation (Figure 3.4C). Within these plots, the returned estimates for the parameter values by the maximum likelihood approach are also shown. The parameter estimates obtained using this approach were obtained through an exhaustive, brute force search of the parameter space. This was done in place of the EM algorithm used in the previous chapter as the convexity of the landscape was not guaranteed. Future iterations of this model could improve upon the computational time this requires as it has now been reasonably demonstrated that there should indeed be an optimal minima for the gradient descent to converge upon. The obtained estimates for the parameter values were subsequently used in conjunction with the Poisson spike train to generate a distribution, Y', from which a new inferred current trace could be created (Figure 3.4B, bottom).

In addition to the computational time cost associated with the current computa-

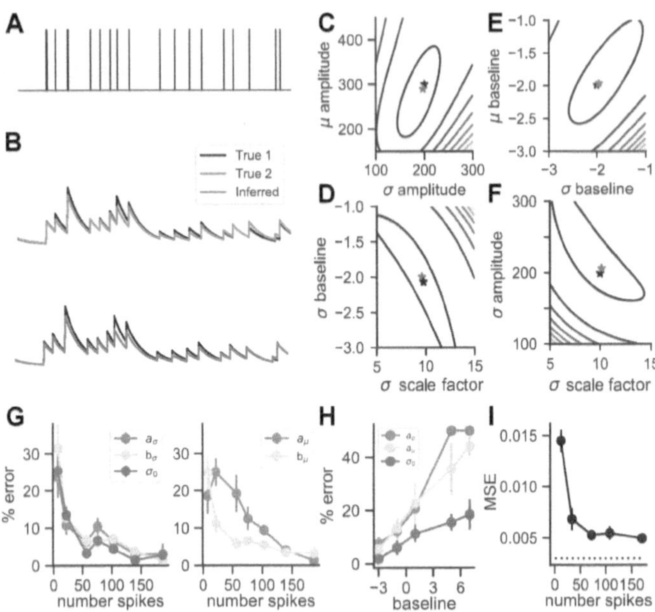

Figure 3.4: **A)** Simulated Poisson spike train marks pre-synaptic stimulation. **B)** Simulated post-synaptic currents of the spike train in **A** for independent samplings of the true parameter set (top) and samplings of true and inferred models (bottom). Independent samplings of the ground truth parameters are taken to show the variance between samplings, as the amplitudes are treated as a random variable. Negative log-likelihood landscape as a function of **C)** μ- and σ-kernel amplitudes, **D)** σ baseline and scaling factor, **E)** μ and σ baseline and **F)** σ scaling and amplitude. **G)** average error of σ parameter (left) and μ parameter (right) given training size. **H)** Average error of amplitude and σ scaling factor as a function of saturation; **I)** Mean square error (MSE) of inferred and true model as a function of training size. Dashed line is MSE between independent samples of the true parameter set. All error bars are s.e.m.

tional state of this method, the limitations and pitfalls of this estimation technique can be further analyzed by varying the number of spikes used in the inference. Varying the spike count for each of the five parameters in θ_μ and θ_σ, it was found that the inferred parameters closely matched to the true parameters used to simulate the responses after 100 to 150 spikes (Figure 3.4G-H). The efficacy parameters required more data than variance parameters; however, for both kernels, the relationship between parameter estimation error and training size (number of spikes) approaches zero as training size gets large. As may be expected from earlier discussion in the chapter, parameters that do not regulate the efficacy lead to poor performance by this inference method. For example, a facilitating efficacy kernel being added to a high baseline parameter leads to saturation in the nonlinear readout, and efficacy facilitation is not observed. In the model, this results in a poor estimation of kernel amplitude on simulated data when the baseline is high (Figure 3.4H). Finally, to test the predictive power of the model a separate artificial Poisson input was generated, and the mean squared error is calculated between the inferred and true model (Figure 3.4I). The prediction error of the inferred model almost matched that of the true model, even when the inference was based on less than 100 spikes. Through these statistical measures, it can be seen that applying the maximum likelihood methodology to the proposed model allows for the successful characterization of simulated stochastic data.

3.4   Discussion

To review briefly, a linear-nonlinear approach, inspired by the success of similar frameworks used for characterizing cellular response [98], has been proposed here as an improved model for characterizing synaptic response properties. And indeed, core elements of subcellular [102], cellular [99, 103–105], and network signaling are successfully captured using the proposed linear-nonlinear framework. The efficacy kernel, in particular, acts as a powerful tool for capturing activity-dependent changes in the synaptic efficacy, including STF, STD, and their combination by simply switching the kernel polarity, although the focus of the experimental examples demonstrated here has been STF. Further, even in its deterministic form, the model successfully reproduces the observed behaviour from extracellular calcium concentration manipulations seen to affect high-frequency stimulation responses, and can naturally capture long-lasting effects such as post-burst facilitation. Importantly, this is done on multiple timescales, a marked improvement on existing linear models [36]. By extending the deterministic form to account for dynamics in a stochastic framework, an automatic characterization metric is elucidated through maximum likelihood techniques. The added flexibility and the efficient inference obtained through these methods are of interest to high through-put characterization of synaptic dynamics [106]. Thus the linear-nonlinear model is an improved characterization method from complex, time-limited, and physiological stimulation patterns.

As described in the second chapter, the maximum likelihood approach taken here is a form of variational inference firmly grounded in Bayesian statistics. As a result of this, we improve upon the inference methods, such as least-mean squares [66], commonly used for parameter extraction in some of the earlier models of synaptic dynamics such as the vesicle depletion and TM models [81, 82]. Compared to popular earlier linear models discussed at the beginning of the chapter, we have introduced a framework that, while mathematically more complex, effectively captures both STF and STD dynamics on multiple timescales without the need for preemptively fitting kernel time constants, therefore providing more flexibility than either past model. Indeed the model proposed here has been demonstrated to handle a far more robust range of synaptic dynamics (e.g. facilitation latency, supra- and sublinearity, STF and STD, etc.) than either the vesicle depletion or use-dependent vesicle models are equipped for. The vesicle depletion model, for example, has been noted as ineffective at capturing slow forms of STD [36]; however, the model proposed here is robust to the multiple timescales and types of STP behaviours observed in synapses. More explicit comparisons were made throughout the work in this chapter between the proposed linear-nonlinear model and the existing TM model, as shown in Figure 3.2. Notably the TM model, unlike the linear-nonlinear model, was unable to effectively capture the different nonlinear behaviours occurring at physiological calcium levels without major changes to its original framework. From this it was shown that the linear-nonlinear framework is better able to characterize the evolving timescales, and nonlinearities present in synaptic dynamics than the existing linear frameworks.

Additionally to the improved abilities to capture behavioural traits of synaptic dynamics, the methods proposed here are also a statistical improvement on the fitting of these dynamics. Indeed the methods proposed here provide an avenue for characterizing the complex dynamics and multiple timescales present *in vivo* for synapses that cannot be easily captured using existing linear models without major modification [23, 36, 92]. For the linear-nonlinear framework here even unknown time constants could be found by fitting the coefficient of a combination of nonlinear basis functions, as has been done for other linear-nonlinear models [98–101]. This improved fitting ability further supports the utility of this approach to characterization over the existing frameworks.

Imperative to understanding the characterization of synaptic dynamics obtained here through the use of two time-dependent functions is questioning what their relative biophysically tractability is. Considering a relationship with neuronal excitability, the exact biophysics are likely unable to be isolated, rather this model should be considered as resulting from a mixture of independent mechanisms. The efficacy kernel, for example, is likely to reflect residual presynaptic calcium concentration, and the changing size of the readily releasable pool [22], but will likely also be reflective of many other possible mechanisms. Likewise, the variance kernel is dependent on membrane resistance and capacitance, but also low-threshold channel density. The relative importance of these processes, however, is untenable with the methodology described here. Despite this reduction in ease of interpretability, the increased flexibility of the proposed linear-nonlinear framework, and exhaustively tiling the time constant space (hence accounting for multiple time scales) makes it a marked improvement on popular existing models, such as the discussed TM model.

Finally having established the utility of the proposed model, and determined its relative level of biophysical tractability, it is worth considering the implications of such a model as successfully capturing synaptic dynamics. Taking a linear-nonlinear approach to modeling short-term dynamics at a cellular level falls in line with previous modeling and experimental work that has established that dendritic integration can follow a hierarchy of linear-nonlinear processing steps [102, 107, 108]. Subcellular compartments filter and sum synaptic inputs through an integration kernel encapsulating local passive and quasi-active properties. Active properties are responsible for a static nonlinear readout and communication in the direction of the cell body. Combined with the approach presented here, this integration model may be extended by one layer where presynaptic spikes first pass through a linear-nonlinear step before entering dendrites. As synapses separately located, or those synapses belonging to different pathways, may have different synaptic dynamics [109, 110], and because spiking neural codes can multiplex streams of information [5, 111, 112], there is then the capacity for these synaptic properties to extract different streams of information from multiple pathways and to process these possibly independent signals in segregated compartments.

Chapter 4

Synaptic transmission and redundancy

4.1   Introduction

Neurons communicate through synaptic release events, a process which has historically been quantified using information theory. The central synapses of interest here have been shown earlier in this thesis to have varied properties and dynamics. These are note entirely intuitive for communication between cells, raising questions of how these traits contribute to information transmission. For synapses, transmitting information between neurons is a key role [16]; as such it is reasonable to assume that the behaviours and structures of synapses should be well set for this task. Whether this benefit is how much information is transferred [80] or if it is an optimization of the energy cost to information transferred [19] remains the subject of debate. Short-term dynamics, as well as pre- and postsynaptic factors (e.g. release stochasticity, amount of vesicles released, type of postsynaptic receptors, etc.) affect the amount of information communicated between cells [78]. As discussed earlier, central synapses display several characteristics that are not intuitively helpful for optimizing information transfer such as: high variability on a response-by-response basis, high noise levels in synapses, sparse connectivity, and a broad synaptic weight distribution [16, 45, 78]. Not all of these properties are unique to central synapses; however, the noisiness of these synapses and their tendency to have seeming release failures has led to much confusion. Recent work has suggested there may be benefits to noise during information transfer [113], and there have been arguments that splitting information transmission over several less reliable, but energetically inexpensive, channels reduces cost requirements [79, 80]. However, in light of the unimodal framework proposed earlier in this thesis, further investigation into what this might mean informationally is important.

In this chapter, we seek to demonstrate preliminary support that central synapses may be optimized for transferring information between neurons in the context of postsynaptic receptor activation. Here a numerical approach is taken to simulate transmission and evaluate the communication between the presynapse and postsynapse. This is done in the context of presynaptic glutamate release and postsynaptic receptor population response.

4.2   Information theory

Information theory is a branch of study in which the quantification, compression, and storage of information are studied. A quantity of interest within this field of study is the mutual information, which has been used in many different fields of study including physics, electrical engineering, neurobiology, and computer science. Mutual information is a measure of the average reduction in uncertainty about a random variable, x, resulting from learning the value of a second random variable, y, or vice versa where the variables x and y are not independent [114]. That is, the mutual information tells us the average amount of information that x conveys about y. This quantity is defined as:

$$I(X;Y) = \sum_{x,y} P(x,y) \log \frac{P(x,y)}{P(x)P(y)}, \tag{4.1}$$

where $I(X;Y) \geq 0$, and equals zero only in the case where x and y are independent [114]. Entropy is closely related to mutual information; indeed mutual information can be written in terms of the difference between the marginal and conditional entropies [114]:

$$I(X;Y) = \sum_{X,Y} P(X,Y) \log \frac{P(X,Y)}{P(X)} - \sum_{X,Y} P(X,Y) \log P(Y)$$

$$I(X;Y) = \sum_{X,Y} P(X)P(Y|X=x)\log\frac{P(X)P(Y|X=x)}{P(X)} - \sum_{X,Y} P(X,Y)\log P(Y)$$

$$I(X;Y) = \sum_{X,Y} P(X)P(Y|X=x)\log P(Y|X=x) - \sum_{X,Y} P(X,Y)\log P(Y)$$

$$I(X;Y) = \sum_{X} P(X)\left(\sum_{Y} P(Y|X=x)\log P(Y|X=x)\right) - \sum_{Y}\left(\sum_{X} P(X,Y)\right)\log P(Y)$$

$$I(X;Y) = -\sum_{X} P(X)H(Y|X=x) - \sum_{Y} P(Y)\log P(Y)$$

$$I(X;Y) = H(Y) - H(Y|X)$$

The mutual information is symmetric, so this result can equivalently be written as $I(X;Y) = H(X) - H(X|Y)$ [114]. Further, the entropy will always be positive, and so $I(X;Y) \leq H(Y)$. Intuitively, the marginal entropy is a measure of the uncertainty of the random variable, and the conditional entropy, $H(Y|X)$, can be thought of as a measure of what the random variable x does not tell us about y. In the framework of neuroscience, mutual information is useful, as it allows us to quantify measures such as the information neurons provide about a given stimulus, or measuring the encoding abilities of a neuron for a given signal or behaviour [115]. This metric provides a critical function as understanding how the brain compresses, encodes, and integrates information is central to understanding brain function.

4.2.1 Information theory and synapses

In addition to its widely used applications at a neuronal level [115], information theory has been applied at the level of synaptic transmission [19, 80]. Research in this framework has included exploration of energy costs in synaptic transmission [19], where a rate-based model measuring the mutual information between incoming spike trains and output EPSC showed a single release site to be more informative than multiple release sites depending on the probability of release. As mentioned in the introduction, in the context of central synapses this research body has focused on the observed unusual qualities of these synapses including the high variability and noise levels between individual release events, transmission failures, the discourse between the univesicular and multivesicular frameworks, and sparse inter-neuron connectivity [16, 19, 80]. Transmission variability and the noisiness of these synapses is not intuitive - one would assume that high variability and noise would detract from successfully transferring information across synapses. However, it has been postulated recently that the presence of noise within synaptic dynamics is a feature of neuronal function rather than a bug [113]. In particular, this proposal is made in the context of the brain evaluating probabilities - for example, evaluating uncertainty to estimate whether a given behaviour (e.g. crossing the street) is a safe behaviour. This uncertainty is proposed to arise in part from the strengths of synaptic connectivity, where synapses evaluate their uncertainty and communicate it through adding noise [113]. This comes after earlier research demonstrated high noise at central synapses to improve memory storage under size constraints [16], and that noise entropy increases information in depressing synapses [80]. Further, release failures have been proposed as a method of disrupting input autocorrelations to reduce the level of cost-inefficient encoding redundancy in neurons [45, 80]. The energetic and efficiency costs of synaptic transmission have played a large role in the information literature, demonstrating features like multivesicular release to be energetically inefficient on time scales long enough to increase synaptic strength [19], and that smaller, more random releases can maximize the efficiency of information transfer [79]. Together this suggests that, despite their counter-intuitive properties, synapses are well made to reduce redundancy and energetic cost while optimizing information transfer.

In other brain regions, it has been shown that more than one type of neuron can encode the same filtered stimulus; however, the different cell types signal the presence of the feature with different thresholds [116]. This seems at-odds with the idea that neurons act to reduce redundancy for optimal information transfer. Extending this work to the synapses considered in this thesis led to questioning the seeming redundancy of having both N-methyl-D-aspartate (NMDA) and α-amino-3-hydroxy-5-methyl-4-isoxazolepropionic acid (AMPA) receptors, both of which populate the post-synapses and respond to glutamate. That is, when a synaptic transmission occurs and glutamate is released, both receptor populations open in response to the exocytosed glutamate with different thresholds of activation. The details of the properties of these receptors, and their interactions with the released glutamate will be elaborated on further below; however, importantly the different thresholds of opening for NMDA and AMPA receptors are such that NMDA responds to lower concentrations of glutamate than AMPA [117]. Despite this, NMDA cannot open without the depolarization resulting from AMPA opening and operates with slower kinetics than AMPA [78]. This seems inconsistent with the idea that synapses are designed for optimal information transmission, and is of particular interest as much of the literature has focused on either the presynaptic aspects of transmission (vesicle releases) [79, 80] or on comparing rate-based distributions of the input and output spikes [19, 113]. Before continuing to investigate this anomaly, it should be asked: is there a case where only one receptor type is active? That is, are their circumstances in which NMDA should detect transmission that AMPA does not? And if there are, what benefits might there be to having one or more reporters for glutamate in different conditions?

4.3   Silent releases

As was proposed in chapter 2, there is the possibility that synaptic transmission occurs in a unimodal framework. This has unique implications for what synaptic release events look like in practice. That is, it fits with the idea of *silent releases* - release events with glutamate amounts are small enough that they fail to trigger a postsynaptically detectable AMPA receptor-mediated response, however, NMDA receptor-mediated response is detectable [1, 118, 119]. This concept is different from postsynaptically silent synapses, which have been well reported in the literature and are defined as synapses where a presynaptic action potential fails to evoke a postsynaptic action potential [35, 118–120]. Silent release responses are different as they imply releases are still successfully occurring, but remain undetected under standard conditions due to the voltage dependency of NMDAR opening, however could be recorded under the correct experimental conditions. This unease of detection makes finding evidence to support the presence of silent *releases* that are not silent *synapses* experimentally a challenging task. Despite the challenges, there has been evidence in the hippocampus that, under strong depolarization conditions, NMDAR mediated currents can be observed with no AMPAR-mediated currents [35, 120]. The unimodal framework of synaptic transmission proposed in chapter 2 suggests that there is statistical support for a model in which silent releases, rather than just silent synapses, are present. In other words, in the unimodal characterization of non-probabilistic release events, there must be small release events that happen below the AMPAR threshold, suggesting theoretical support of silent releases.

If releases occur silently, or below the AMPA detection threshold, we must then ask: How prevalent are they? And, if there are silent releases, what benefit might they provide? Historically, response amplitudes for 'failed' release events in AMPAR and NMDAR mediated responses have been determined using a threshold-based method. Typically this is done by setting a threshold around twice the standard deviation of the noise in the AMPA-mediated response measurements, below which all responses are considered failures. However, as this threshold is set based on the success and failures of AMPAR-mediated responses, there could be a meaningful measure of successful NMDAR-mediated responses being misclassified as failures. There will also be some amount of mislabelling of successes and failures in the AMPA responses using this metric, as is an inherent bias in applying a hard threshold. An example of this is shown in Figure 4.1A. If this thresholding indeed overestimates the failures of NMDA mediated responses, performing parameter estimation on the response distributions of NMDA and AMPA should show the probability of failure of AMPA to be significantly higher than that of NMDA. This result would indicate that setting a threshold for success via the AMPA failures will under-count NMDA successes. This analysis was done using the bimodal model framework established in chapter 2 on a set of minimal-stimulation voltage-clamp EPSC data for NMDAR-mediated (voltage clamp at 40mV) and AMPAR-mediated (voltage clamp at -70mV) responses from 5 neurons (Example of one neuron shown in Figure 4.1A). The resulting estimation of the failure probability by applying the EM algorithm to this data agreed with the prediction that NMDAR-mediated responses had a significantly lower failure probability (MannWhitneyU Test: $p < 0.05$; Figure 4.1B). Together this corroborates the idea that release events occur below the AMPA detection threshold.

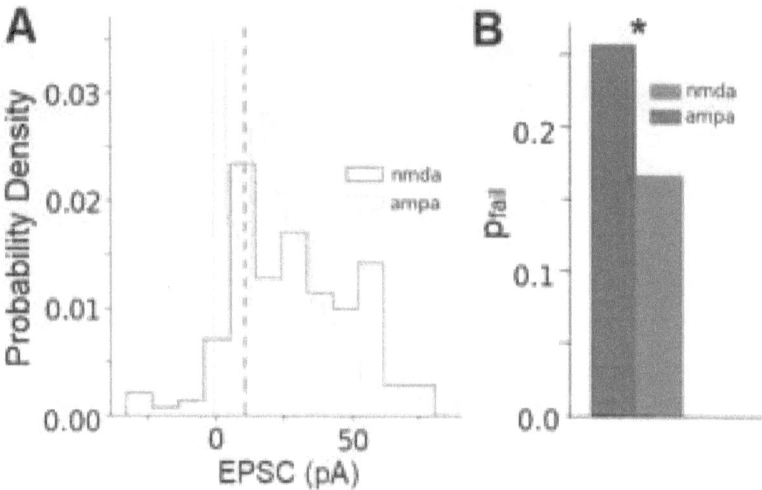

Figure 4.1: **A)** EPSC histograms for minimal stimulation NMDA and AMPA currents from one examplar neuron (experiments performed by Dr.Anup Pillail from the Béïque lab). A threshold (brown dashed line) is placed, based on the noise in the recording for the AMPA EPSC, at $2\sigma_{noise}$. Below the threshold the current values are counted as 'failures'. **B)** Comparison of probability of failure for NMDA and AMPA EPSCs predicted by expectation-maximization algorithm using bimodal non-convolved model.

Importantly, silent releases, and by extension the unimodal framework, change the theoretical cause of detectable 'silence' in a synapse from a postsynaptic cause (i.e. the lack of AMPA receptors observed in silent synapses) to a presynaptic one (i.e. too small of a release to detect). The suggestion that detectable 'silence' in synaptic response derives from a presynaptic effect has been suggested before as an effect of low release probability [121] or spillover from nearby synapses [118]. However, this differs from the proposal here where, under the unimodal framework, release size rather than release probability would be the cause of presynaptically silent releases. This makes silence a much more transient quantity as an increase in release size would be sufficient to 'unsilence' the synapse. Conversely, silent synapses, are postsynaptically effected; that is, to unsilence silent *synapses* there must be an insertion of AMPA receptors into the postsynapse [35, 120]. No presynaptic increase in release will unsilence these synapses. Together these preliminary outcomes suggest that there may be a meaningful role in transmission for the difference in detection thresholds of NMDA and AMPA receptors.

4.3.1 NMDA and AMPA receptors

Having established that there is a seeming redundancy to having two glutamate detectors with different thresholds for synaptic release, and that there is potentially a meaningful cause for this difference, we will now discuss these receptors in further detail. Synaptic transmission events require postsynaptic receptors to allow entry of the desired neurotransmitter to the receiving cell. NMDA and AMPA receptors are both found in the postsynaptic membranes of central synapses and have been largely accepted to be relatively co-localized on dendritic spines, simultaneously responding to glutamate release [118, 120, 122]. AMPA recepetors require a higher concentration of glutamate to bind than NMDA in order for the receptor to open. Once the AM-PAR are open, and hence allowing current to pass, the channel may undergo rapid depolarization (thus stopping the current). For AMPAR this happens on the order of $1ms$ [118, 123]. Conversely, at resting membrane potential, the NMDA channel is blocked by a $Mg2+$ ion; unblocking this is non-instantaneous [124]. To unblock the NMDAR, the postsynaptic cell must be sufficiently depolarized; this occurs on the scale of milliseconds [124]. This depolarization arises with the opening of the AMPAR's in physiological conditions. That is, in a response the AMPAR receptors, despite requiring more glutamate concentration to open, precede NMDAR's opening [118, 123, 124].

NMDAR have been shown to be less variable in their responses than AMPAR [122], with Hill coefficients - a measure of their binding cooperativity - such that NMDAR is in the range of 1.7 to 2.1 [117]. These binding cooperativity values are higher than those observed in AMPAR (1.3 to 1.6) [117]. This difference in binding cooperativity is consequentially why NMDAR are capable of detecting smaller concentrations of glutamate. However, due to the required membrane depolarization, NMDAR will not open under normal depolarization conditions if AMPAR does not open. There are synapses in the hippocampus that have been reported to have responses entirely generated by NMDAR-mediated activity [118, 120], which would qualify as a 'silent synapse' as the voltage dependence would be effectively silent at resting membrane potentials. Further, immunoreactivity studies have demonstrated NMDA to be much more frequently present postsynaptically than AMPA receptors [122]. However, if there are AMPARs present, then the NMDAR-mediated response may be a silent release.

The frequent presence of both AMPAR and NMDAR receptors on the postsynapse suggests that there should be an informative reason to have both receptor types present. However, if silent releases are occurring, this should be supported by an information-based benefit. That is, there should regularly be a benefit to having more than one glutamate detector present, but there should also be a paradigm where having a single low concentration detector provides more information than having two. To explore this, I develop a numerical framework to simulate the presynaptic release distributions of glutamate molecules using the previously discussed unimodal framework and subsequently find the relative receptor activation distributions for NMDA and AMPA. The cases of only an NMDA receptor population and both an NMDA and AMPA populations are considered in context of being better or worse for communication between cells. Importantly, this is only a preliminary metric to demonstrate the presence of two glutamate receptors in the postsynapse to not be a redundancy.

4.4 Simulating synaptic transmission

To consider the communication across the synapse requires a numerical metric to link the two sides of the synapse. Here this is achieved through relating the presynaptic glutamate release to the postsynaptic receptor activation. This relationship can be attained from the Hill equation [39], such that:

$$R = \frac{\left(\frac{G}{K_A}\right)^{\eta}}{1 + \left(\frac{G}{K_A}\right)^{\eta}}, \tag{4.2}$$

where G is the number of glutamate molecules, η is the Hill coefficient, K_A is the number of molecules at which the half-maximal response is reached (also sometimes called the EC_{50} value) and R is the receptor activation. Having this relationship it is then possible to determine a distribution of receptor activation from the distribution of glutamate, which we have already formulated in chapter 2 as gamma-distributed. While initially attempts were made to arrive at an analytical solution to this by performing a random variable transformation, this proved intractable to a convincing level. As such, a numerical approach to this problem is taken. Specifically, a distribution of glutamate, $P(G)$, is formulated with the average number of glutamate molecules, G, from the release of a single vesicle taken from the literature: $G \approx 3000$ molecules, with a reported standard deviation of 700 molecules [117,125]. Using this value, the parameters for the gamma distribution (γ and λ) are found using the mean and standard deviation relations described in chapter 2. Having this, a simulated version of the unimodal model of glutamate release is made (Figure 4.2A), where the number of vesicles, n, is randomly chosen per release trial such that $n \in [1 : 10]$. To account for the earlier observed estimates of n as generally $n \leq 5$, the range of n values was weighted such that $n \in [1 : 5]$ is more likely than $n \in [6 : 10]$. The resulting distribution of glutamate molecules, $P(G)$, is then:

$$P(G) = \sum_{k=1}^{n} \frac{e^{-G/\lambda_G} G^{k\gamma_G - 1}}{\Gamma(k\gamma_G)\lambda_G^{k\gamma_G}}, \tag{4.3}$$

where γ_G and λ_G refer to the parameter values determined based on the average number of glutamate molecules in one vesicle. From this distribution of glutamate molecules, equation 4.2 can be used to obtain the corresponding receptor response distributions for NMDA and AMPA using the appropriate η and K_A values for each. For the simulations here the values of K_A and ranges of η for both NMDA and AMPA used to simulate the receptor response distributions are taken from the results of Pankratov and Krishtal [117]. The simulated data generated here is such that a glutamate distribution is sampled as a single release event (of n vesicles) for a total number of release events, or trials, up to N_{trials}. Specifically, for a given trial, a glutamate release distribution is generated by sampling from the gamma-mixture distribution for whatever n was chosen. The number of glutamate molecules in this release, G, is then used to calculate the postsynaptic receptor activation R. This is then repeated for N_{trials} to create the distribution of receptor activation, which is necessarily conditioned on the amount of glutamate that was released, and so we have $P(R^{(j)}|G)$, where $j = \{NMDA, AMPA\}$. The values of G and resulting $R^{(j)}$ are stored in an $i \times h$ matrix, where the rows, i, contain the number of glutamate molecules released and the fraction of receptor activation for each trial $i \in \{1, ..., N_{trials}\}$, and there are $h = 3$ columns, one each for the glutamate released, and the fraction of NMDA and AMPA receptors activated, respectively.

To verify the accuracy of this simulated method, the resulting receptor activation distribution can be used to create simulated EPSC distributions by multiplying the probability of receptor activation by the appropriate voltage (V_{NMDA}: 40 mV and V_{AMPA}: -70 mV) and unitary conductance (g_{NMDA}: 40 pS and g_{AMPA}: 15 pS) [126] values for AMPA and NMDA and adding noise. This was done here (see Figure 4.2B, right) and could then be compared to the minimal stimulation EPSC distributions obtained experimentally by Dr. Anup Pillail from the Béïque lab for NMDA and AMPA (example from one neuron in Figure 4.2B, left). The distribution of NMDA generated in Figure 4.2B is more right shifted than that of the example EPSC, however the example EPSC has a higher maximum current value, thus the means are almost the same ($\mu_{sim.}^N = 11.5pA; \mu_{rec.}^N = 10.2pA$). This is likewise the case for the simulated versus recorded AMPA currents ($\mu_{sim.}^A = 10.3pA; \mu_{rec.}^A = 11pA$). This reasonable agreement suggests the metric of determining the receptor activation is effective.

Using the Hill equation to relate glutamate release and receptor activation, it is helpful to have some intuition of the properties of this equation. An important characteristic of this relationship is that the rate at which R saturates is based largely on the value of η, which varies depending on the receptor type - specifically, η controls the slope of the curve. As mentioned prior, η may be thought of physically as a measure of the cooperativity of the binding process [39]. A Hill coefficient of 1 indicates independent binding and values greater than 1 indicating positive binding cooperativity (Figure 4.2C). There is a theoretical upper limit of the Hill coefficient when binding is completely cooperative (all binding sites would bind glutamate simultaneously); however, this limit is never reached in practice [39]. In the case of the receptors of interest here, the Hill coefficient of AMPAR's is lower than that of NMDAR's, hence NMDA receptors experience more cooperative binding. The K_A, being the required number of molecules for half max activation, will shift the point at which the slope starts. For the cases considered here, the K_A's are fixed values for NMDA and AMPA with $K_A^{(AMPA)} > K_A^{(NMDA)}$, so the slope of NMDA will increase at lower values of G. Having a metric to simulate release events at central synapses and the corresponding receptor activation, it is then possible to pursue our goal of demonstrating the presence of two receptor populations as potentially useful.

4.4.1 Communication between pre- and postsynapse

As the aim of the work here is solely a preliminary foray into showing that the two receptor populations on the postsynapse are redundant, an information theory approach is not strictly used. Rather, elements of information theory are used to devise a metric for examining the effective transmission detected by having two different detection thresholds on the receiving cell. Here effective transmission refers to a metric of quantifying, preliminarily, the successful communication of the presynaptic signal (glutamate release) to the postsynapse (receptor activation). The details of how this is achieved are detailed as follows.

For the relationship of interest here, we will begin by considering the receptor's interactions with the release neurotransmitter individually. The mutual information between presynaptic release and receptor activation can be written as:

$$I(G; R^{(j)}) = -\sum_{r \in R} \left(\sum_{g \in G} P(g)P(r^{(j)}|g) \right) \log \left(\sum_{g \in G} P(g)P(r^{(j)}|g) \right) \\ + \sum_{g \in G} P(g) \sum_{r \in R} P(r^{(j)}|g)\log P(r^{(j)}|g) \tag{4.4}$$

where r and g are instances of R and G, the superscript $(j) = \{NMDA, AMPA\}$ indicates the type of receptor, and hence the $K_A^{(j)}$ and Hill coefficients,$\eta^{(j)}$ used. For the question of interest here this quantity will be used to describe the effective transmission between the presynapse and *one* of the postsynaptic receptor populations. Importantly, as we wish to consider both populations we must take the sum of these values. However, in a synapse both receptor populations will interact with the same presynaptically released glutamate at the same time. Intuitively, this indicates that a value must be subtracted from our summation to avoid double counting the glutamate interactions. For simplicity the correlation coefficient, $\rho_{N,A}$, was approximated here as accounting for that overlap, and is hence subtracted off the two receptor case terms, $I(G; R^{(NMDA)})$ and $I(G; R^{(AMPA)})$. The equation that results from this is:

$$ET_{Total} = ET_{NMDA} + ET_{AMPA} - \rho_{N,A} \tag{4.5}$$

where ET_{NMDA} and ET_{AMPA} are $I(G; R^{(NMDA)})$ and $I(G; R^{(AMPA)})$, respectively, as calculated from equation 4.4, and ET_{Total} is the effective transmission.

For the case of a single receptor population, the second Hill coefficient $(\eta^{(AMPA)})$ is fixed such that $P(r^{(AMPA)}|g) = 0$ for all $g \in G$, which will yield a zero value for ET_{AMPA}, and a zero correlation coefficent, hence: $ET_{Total} = ET_{NMDA}$ when there is only one receptor type.

Performing simulations to attain the relevant distributions as described in the previous section, the effective transmission from equation 4.5 was calculated at different values of $\eta^{(j)}$ within the known ranges for AMPA and NMDA [117]. The resulting average is then plotted as a function of the difference between $\eta^{(NMDA)}$ and $\eta^{(AMPA)}$ in Figure 4.2D, where a difference, $\Delta\eta$, of zero corresponds to the case of a single Hill coefficient, and hence a single receptor population. From this plot it can be seen that the maximum for each curve, regardless of the Hill coefficient (η_2) for the NMDA receptor

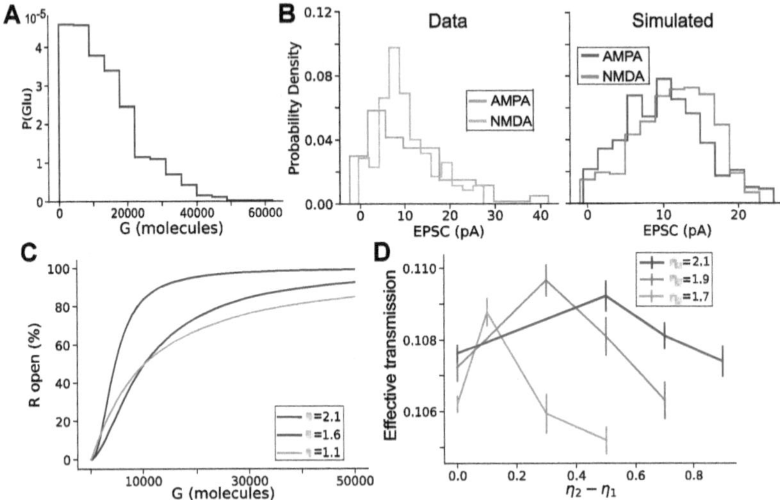

Figure 4.2: **A)** Example simulated glutamate molecule distribution; **B)** Comparison of recorded (left) and simulated (right) EPSC distributions for NMDA and AMPA. EPSC for AMPA plotted as $\times(-1)$ for visual comparison. **C)** Receptor (R) activation from glutamate molecules (G) as given by Hill equation for different Hill coefficients, η; **D)** Effective transmission of two receptor populations ($\eta_2 - \eta_1 > 0$) and a single receptor population ($\eta_2 - \eta_1 = 0$).

population, is at a point where $\Delta\eta > 0$. That is, a point where both receptor populations are active. Based on the metric used here then, there is preliminarily an incentive for having more than one receptor population in the postsynapse. Of further note, the minimum for each curve is not at $\Delta\eta = 0$, where only one receptor population is considered. This is of interest as it further supports the unimodal framework proposed. That is, it supports that there may be a set of circumstances in which transmission is more favourably done as small releases below the postsynaptically detectable threshold. This observation would also fall in line with the aforementioned silent release framework.

4.5   Discussion

In this chapter the objective was to investigate the seeming redundancy of there being two glutamate receptors in the postsynapse. Specifically, the question was raised as to whether it is a redundancy at all, or if there is evidence to support that it may be beneficial. To interrogate this possibility here, a numerical framework was created to simulate synaptic transmission and, using the Hill equation, explore the effective transmission that occurs. Notably, this also provides a framework for theory on silent releases to begin to be investigated. As was noted in chapter 2, silent releases have important implications in modeling transmission because their presence would further support the statistically validated unimodal framework of release proposed in that chapter. That is a framework in which the variability of successful response or 'silence' arises due to the size of release and not failure to release. This should not be assumed to mean there could never be failures, rather it proposes that if there are true failures they occur with a substantially lower frequency than has been assumed in the literature thus far. The occurrence of silent releases would then create two versions of synaptic silence, the better documented postsynaptic type associated with silent synapses, but also a presynaptic silence that could be resolved by increasing the release size [118, 119, 121]. In the results of this chapter, the unimodal framework becomes framed in terms of single receptor population activation. That is, the case where a release is too small for AMPA to detect, but would be detectable to NMDA. The support for silent releases here begins rather simplistically through verifying that, based on a bimodal failure-release model, the probability of failure of NMDA-mediated EPSCs is significantly lower than the failure probability of AMPA-mediated EPSCs (Figure 4.1B). Thus, when thresholds based on AMPA failures are used to classify successes and failures of both receptor classes, there will be misclassified NMDA responses. Those NMDA responses that are misclassified would then be silent releases as AMPAR are present postsynaptically [119]. Having further support for the unimodal framework suggests there to be a benefit to having small releases. Whether these small releases are the result of 'kiss-and-run' releases (see Chapter 1) or smaller than average vesicles, or from spillover of nearby synapses remains unknown.

The support of silent releases, in addition to changing the framework in which synaptic dynamics are discussed, provides a potential physiological reason for the presence of two glutamate receptors. As described earlier in this chapter, it is reasonable to assume there should be some benefit to having two glutamate detectors under normal successful release conditions. This assumption is further supported by the history of central synapses having benefits for each of their seemingly counter-intuitive characteristics [19, 45, 80, 113, 118]. Additionally it has been shown in other regions of the brain that there are, under certain conditions, benefits to having two detectors with different thresholds for an incoming stimulus [116]. The preliminary results here offer support that this may indeed also be the case in terms of postsynaptic receptors NMDA and AMPA (Figure 4.2D). Based on the results here, regardless of what the Hill coefficient for the NMDA receptor population is, the most effective transmission ought to occurs when both receptor populations are active. This fits well with the idea that synapses may be well set up to optimize communication.

Interestingly, the minimum of the curves in each case was not when only NMDA was active, but rather when there was the greatest difference between the Hill coefficients of NMDA and AMPA (see Figure 4.2D). More specifically, the minimum occurs as the difference between the two coefficients gets larger ($\eta^{(AMPA)} \to 1$). While exact inferences should not be made on the metric used here, this decrease may be a result of lower Hill coefficients for AMPA approaching independent binding. This suggests the existence of a regime where a single glutamate receptor is better than having two. More broadly, this suggests a single reporter for glutamate would be better in cases where the second reporter is not a particularly good detector. Such a regime could indeed be the case in the lower range of binding cooperativity for AMPA, which approaches independent binding. This would fit with previous evidence that synaptic information transfer acts to reduce redundancy [45]. As, while having two detectors where one is very unlikely to be successful in detection would seem a poor use of resources, if there is not a benefit under some conditions to having two detectors it would also be redundant to have them.

Of note, these information results are preliminary and should not be over-interpreted; this numerical metric of calculation is not as robust as a mutual information calculation. Additionally, this metric does not currently change the noise level between trials or in the results. That is, the noisiness is kept at a constant level in the results presented here. This is important when comparing with existing research, much of which has postulated the signal-to-noise ratio, and the level of noise as important [80, 113, 116]. Previous research in other neurons has shown that higher noise regimes can lead to information being maximized by a case of a single detection threshold where two are available [116]. It could be that this would also be the case for the glutamate receptors considered here, as high noise would make successful signal detection harder and so a single threshold would have an easier time discerning a signal than a more fine-grained dual detection. However, such conclusions would require further development of the numerical framework, most likely into a more formal mutual information calculation and with more robust simulations. Further, the approach used here does not consider time scales of the releases, making the information calculations approximately those of independent single release events. In a framework where time is considered, high-frequency stimulations, which minimize the amount of time for clearance from the previous release, should be less likely to have cases where small amounts of glutamate are present in the synapse. Despite the simplicities present in this framework, the support for the practicality of having two postsynaptic glutamate receptor types with different thresholds of detection remains a promising avenue of research.

The implications of this research suggesting unimodality as a useful framework within synaptic transmission have far-reaching consequences for both the conditions presented in this chapter, but also the characterizations of central synapses as a whole. The evidence presented in this thesis creates a compelling argument for the utility of unimodal frameworks for the characterization of synapses and capturing cellular level short-term dynamics, as well as reviewing physiological explanations of why and how such seeming failures might happen and circumstances under which they may be more informative than a larger release. Much more work needs to be done to fully investigate the breadth of utility for this overarching framework. Nevertheless, the applicability of this paradigm should not be overlooked as it may well provide meaningful insight into the underlying behaviours and mechanisms of the higher order processes in the brain.

Conclusion

Over the course of this thesis, the synaptic dynamics of hippocampal synapses have been investigated with particular focus on how the properties of these synapses affect information transmission. The existing, extensive research into central synapses has suggested the anti-correlation between release probability and multivesicular release may indicate a regulatory role for MVR in scaling up and down with the probability to avoid oversaturation of the postsynaptic receptors [10]. In line with this, the NCGM model for single synapse release events proposed in chapter 2 demonstrates average vesicle size to be anti-correlated with the number of vesicles releasing. That is, the predictions from the NCGM model suggest that, rather than multivesicularity alone, the synapses optimize transmission toward an average release size. Two outcomes arise from this: 1) vesicle size, which has been missing in many earlier models of synaptic transmission [10,40,49,50], is an important quantity to consider; and 2) synapses avoid over-saturation of post-synaptic receptors, which would have no informative benefit, by controlling the overall volume of neurotransmitter released into the cleft. These two results are further supported by the statistical evidence that a unimodal framework - where probability of release is not a factor - is a comparably good model. Together with the NCGM model, this is a framework shift from modeling hippocampal synapse releases as bimodal events controlled dominantly by release probability and number of vesicles, to events that scale for a particular release size where some may be too small to easily detect postsynaptically.

Adding to the validity of the unimodal framework, it was demonstrated in Chapter 3 that a kernel, parameterized by the mean and standard deviation of the PSC amplitude distribution, can be used to capture the PSC amplitudes and their evolving dynamics. While this model of synaptic dynamics is less intuitive for interpretation than popular existing models, such as the Tsodyks-Markram model, it accounts for import factors that are missing from these earlier models. Notably among these is the linear-nonlinear models accounting for the multiple time scales on which synaptic dynamics evolve and capturing the supralinear dynamics the TM model does not effectively characterize.

Having investigated the physical aspects from a presynaptic (synaptic release) standpoint, in the final chapter of this thesis the postsynapse was considered. A numerical metric was used to probe the seeming redundancy of having two glutamate detectors with variable thresholds. From this it was shown, preliminarily, that there may be a benefit to maintaining both receptor types, suggesting synapses to be generally well equipped to communicate between each other. Further, the proposed framework suggests there may be a regime in which silent releases would be present, further validating the unimodal description of releases as size-based. Together, the work in this thesis paints a picture of synapses as being well equipped for efficient inter-neuron transmission.

Appendix A

Consider the gamma-Gaussian model described by Equation 2.9; to determine CV^2 first note that the mean of Equation 2.9 is $\mu_x = np\gamma\lambda$, the mean of the number of vesicles is $\sum_{k=0}^{n} kb(k|n,p) = np$ and the mean amplitude per vesicle is $\int xg(x|\gamma,\lambda)dx = \gamma\lambda$. The variance of x is then:

$$\sigma_x^2 = (1-p)^n \int (x - \mu_x)^2 G\left(x \mid 0, \sigma_{opt}^2\right) dx \\ + \sum_{k=1}^{n} b(k \mid n, p) \int_0^{\infty} (x - \mu_x)^2 g(x \mid k\gamma, \lambda)dx \tag{A.1}$$

The first term of this equation becomes:

$$(1-p)^n \left(\sigma_{opt}^2 + \mu_x^2\right) \tag{A.2}$$

and the second term is evaluated by centering the quadratic part of the integrand around the mean of a component, $k\gamma\lambda$. This gives for the second term of Equation A.1.

$$\sum_{k=1}^{n} b(k \mid n, p) \int_0^{\infty} \left[(x - k\gamma\lambda)^2 + \gamma^2\lambda^2(k - np)^2 \right. \\ \left. + 2(x - k\gamma\lambda)(k - np)\gamma\lambda\right]g(x \mid k\gamma, \lambda)dx \tag{A.3}$$

To evaluate this expression, one can isolate the variance of the number of vesicles: $\sum_{k=0}^{n}(k - np)^2 b(k|n,p) = np(1-p)$, as well as the variance of amplitude from k vesicles

$\int (x - k\gamma\lambda)2g(x|k\gamma, \lambda)dx = \gamma\lambda^2$. Since the last term vanishes in Equation A.3, yielding:

$$\sigma_x^2 = (1 - p)^n \sigma_{opt}^2 + np\gamma\lambda^2 + \gamma^2\lambda^2 np(1 - p) \tag{A.4}$$

Using $CV_{UVR}^2 = 1/\gamma$ and $CV_{bin}^2 = (1 - p)/np$, taking the ratio $CV^2 = \sigma_x^2/\mu_x^2$, the total CV^2 can then be written:

$$CV^2 = \frac{\sigma_{opt}^2(1 - p)^n}{(np\gamma\lambda)^2} + \frac{1}{np}CV_{UVR}^2 + CV_{bin}^2 \tag{A.5}$$

Appendix B

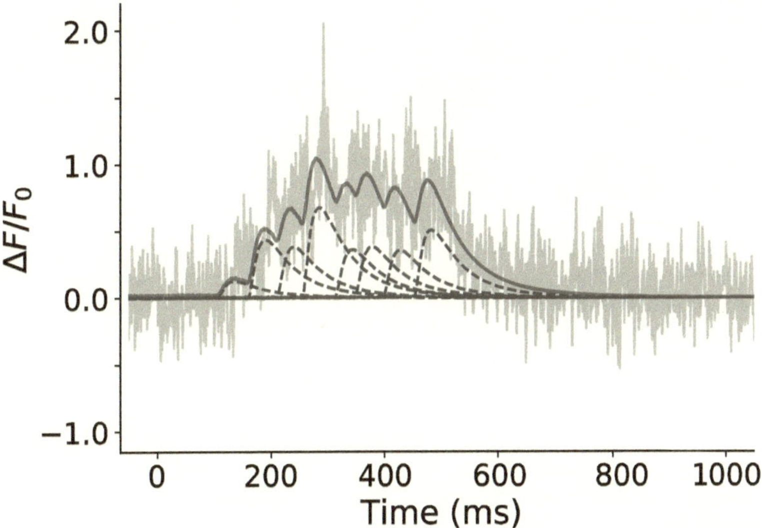

Figure B.1: To account for amplitude effects from the decay of the previous stimuli when extracting amplitudes, multilinear regression was performed. An example fitting using multilinear regression on a $20Hz$ train with eight stimulations ($N_{spine} = 1$) with known *a priori* stimulation times. The red trace is overall fit, and the blue dashed curves are component fits from which the amplitude of each stimulation is extracted.

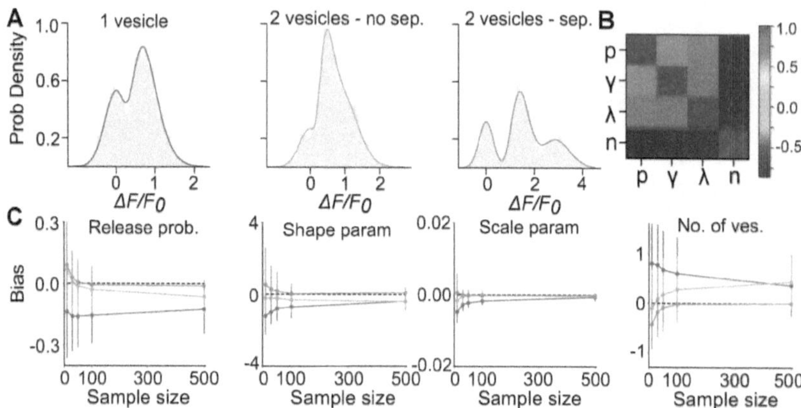

Figure B.2: Prior to use the GGM model was tested for bias in parameter estimates under several simulated distribution conditions. **A)** Count histograms for simulated data and best fit probability density function (full line) for a gamma mixture with left: n = 1 vesicles, a skew $\gamma = 7$, scale $\lambda = 0.12$ and release probability p = 0.6; center: n = 2 vesicles, skew $\gamma = 6$, scale $\lambda = 0.1$ and release probability p = 0.55; and right: n = 2 vesicles, skew $\gamma = 15$, scale $\lambda = 0.1$ and release probability p = 0.51. **B)** Correlation coefficient between parameter estimates of simulated data **A** center, with sample size = 100. **C)** The bias of estimates for left: release probability (p), center-left: shape parameter (γ), center-right: scale parameter (λ), and right: number of vesicles (n) as a function of number of samples. Error bars show parameter estimates standard deviation.

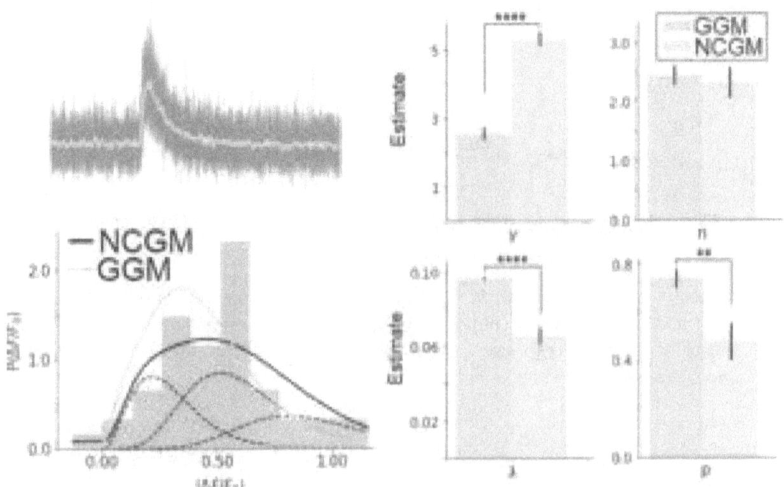

Figure B.3: Changing the method of modeling noise by going from the GGM to NCGM, a model comparison is conducted. Left: (top) Single simulation traces (grey) and their mean (yellow) and (bottom) extracted amplitude distribution fit for both non-convolutional and convolutional forms of model. Right: Parameter estimates for non-convolutional (dark yellow) and convolutional (light yellow) models. Error bars are s.e.m; significance by MannWhitneyU test.

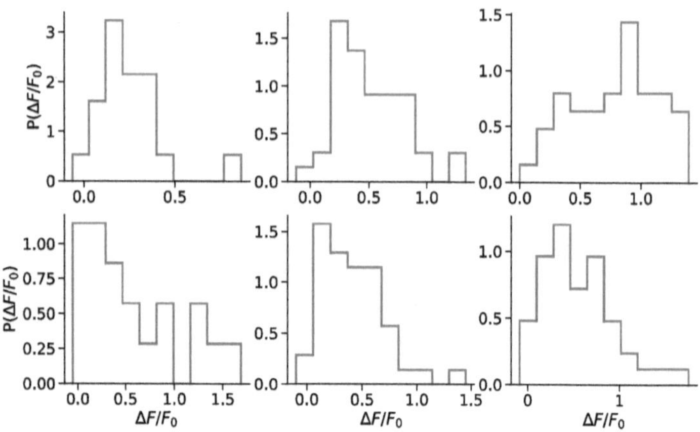

Figure B.4: The inspiration for the unimodal approach to modeling came from the observation that there is not a clear bimodality in the response distributions. Here a subset of example response amplitude distributions from single stimulation protocols under control conditions is shown.

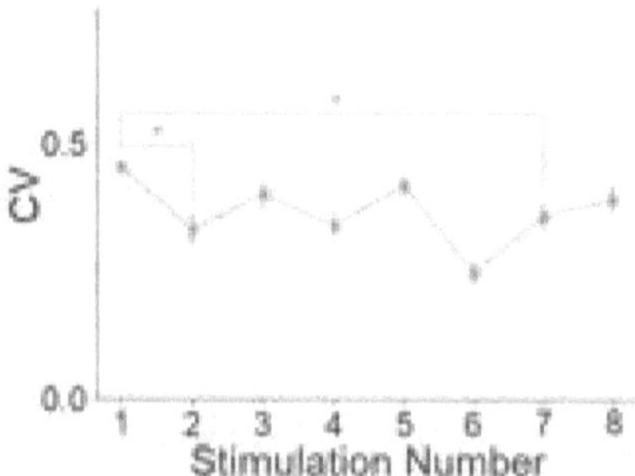

Figure B.5: The coefficient of variation (CV) for the control and TTX conditions of paired pulse stimulation is shown in Figure 2.5C. The mean coefficient of variation ($N_{spines} = 7$) in response per stimulation at $20Hz$ under the control conditions is shown here. Error bars are s.e.m. Significance by MannWhitneyU test.